Workshop on Electromagnetic Inverse Problems 2009

Journal of Physics: Conference Series Volume 255

Manchester, United Kingdom
15 – 18 June 2009

Editors:

Oliver Dorn
Bill Lionheart

ISBN: 978-1-61782-245-2
ISSN: 1742-6588

Printed from e-media with permission by:

Curran Associates, Inc.
57 Morehouse Lane
Red Hook, NY 12571

Some format issues inherent in the e-media version may also appear in this print version.

Printed by Curran Associates, Inc. (2011)

For permission requests, please contact the Institute of Physics
at the address below.

Institute of Physics
Dirac House, Temple Back
Bristol BS1 6BE UK

Phone: 44 1 17 929 7481
Fax: 44 1 17 920 0979

techtracking@iop.org

Additional copies of this publication are available from:

Curran Associates, Inc.
57 Morehouse Lane
Red Hook, NY 12571 USA
Phone: 845-758-0400
Fax: 845-758-2634
Email: curran@proceedings.com
Web: www.proceedings.com

TABLE OF CONTENTS

INTRODUCTION

This proceeding combines selected contributions from participants of the Workshop on Electromagnetic Inverse Problems which was hosted by the University of Manchester in June 2009. The workshop was organized by the two guest editors of this conference proceeding and ran in parallel to the 10th International Conference on Electrical Impedance Tomography, which was guided by Bill Lionheart, Richard Bayford, and Eung Je Woo. Both events shared plenary talks and several selected sessions.

One reason for combining these two events was the goal of bringing together scientists from various related disciplines who normally might not attend the same conferences, and to enhance discussions between these different groups. So, for example, one day of the workshop was dedicated to the broader area of geophysical inverse problems (including inverse problems in petroleum engineering), where participants from the EIT community and from the medical imaging community were also encouraged to participate, with great success. Other sessions concentrated on microwave medical imaging, on inverse scattering, or on eddy current imaging, with active feedback also from geophysically oriented scientists. Furthermore, several talks addressed such diverse topics as optical tomography, photoacoustic tomography, time reversal, or electrosensing fish.

As a result of the workshop, speakers were invited to contribute extended papers to this conference proceeding. All submissions were thoroughly reviewed and, after a thoughtful revision by the authors, combined in this proceeding. The resulting set of six papers presenting the work of in total 22 authors from 5 different countries provides a very interesting overview of several of the themes which were represented at the workshop. These can be divided into two important categories, namely (i) modelling and (ii) data inversion. The first three papers of this selection, as outlined below, focus more on modelling aspects, being an essential component of any successful inversion, whereas the other three papers discuss novel inversion techniques for specific applications.

In the first contribution, with the title *A Novel Simplified Mathematical Model for Antennas used in Medical Imaging Applications*, the authors M J Fernando, M Elsdon, K Busawon and D Smith discuss a new technique for modelling the current across a monopole antenna from which the radiation fields of the antenna can be calculated very efficiently in specific medical imaging applications. This new technique is then tested on two examples, a quarter wavelength and a three quarter wavelength monopole antenna.

The next contribution, with the title *An investigation into the use of a mixture model for simulating the electrical properties of soil with varying effective saturation levels for sub-soil imaging using ECT* by R R Hayes, P A Newill, F J W Podd, T A York, B D Grieve and O Dorn, considers the development of a new visualization tool for monitoring soil moisture content surrounding certain seed breeder plants. An electrical capacitance tomography technique is employed for verifying how efficiently each plant utilises the water and nutrients available in the surrounding soil. The goal of this study is to help in developing and identifying new drought tolerant food crops.

In the third contribution *Combination of Maximin and Kriging Prediction Methods for Eddy-Current Testing Database Generation* by S Bilicz, M Lambert, E Vazquez and S Gyimóthy, a novel database generation technique is proposed for its use in solving inverse eddy-current testing problems. For avoiding expensive repeated forward simulations during the creation of this database, a kriging interpolation technique is employed for filling uniformly the data output space with sample points. Mathematically this is achieved by using a maximin formalism.

The paper *2.5D inversion of CSEM data in a vertically anisotropic earth* by C Ramananjaona and L MacGregor considers controlled-source electromagnetic techniques for imaging the earth in a marine environment. It focuses in particular on taking into account anisotropy effects in the inversion. Results of this technique are demonstrated from simulated and from real field data.

Furthermore, in the contribution *Multiple level-sets for elliptic Cauchy problems in three-dimensional domains* by A Leitão and M Marques Alves the authors consider a TV-H^1 regularization technique for multiple level-set inversion of elliptic Cauchy problems. Generalized minimizers are defined and convergence and stability results are provided for this method, in addition to several numerical experiments.

Finally, in the paper *Development of in-vivo fluorescence imaging with the matrix-free method*, the authors A Zacharopoulos, A Garofalakis, J Ripoll and S Arridge address a recently developed non-contact fluorescence molecular tomography technique where the use of non-contact acquisition systems poses new challenges on computational efficiency during data processing. The matrix-free method is designed to reduce computational cost and memory requirements during the inversion. Reconstructions from a simulated mouse phantom are provided for demonstrating the performance of the proposed technique in realistic scenarios.

We hope that this selection of strong and thought-provoking papers will help stimulating further cross-disciplinary research in the spirit of the workshop. We thank all authors for providing us with this excellent set of high-quality contributions. We also thank EPSRC for having provided funding for the workshop under grant EP/G065047/1.

Oliver Dorn, Bill Lionheart
School of Mathematics, University of Manchester, Alan Turing Building, Oxford Rd Manchester, M13 9PL, UK
E-mail: oliver.dorn@manchester.ac.uk, bill.lionheart@manchester.ac.uk
Guest Editors

A Novel Simplified Mathematical Model for Antennas used in Medical Imaging Applications

M J Fernando, M Elsdon, K Busawon and D Smith
School of Computing, Engineering & Information Sciences, Ellison Bldg,
University of Northumbria, Newcastle NE1 8ST, UK

Email: michael.fernando@northumbria.ac.uk

Abstract

In this paper a new technique is proposed to model the current across a monopole antenna and thereby the radiation fields of the antenna can be calculated. Generally, the Method of Moments (MOM) technique is used for this purpose whereby the integral equations are discretised to find the fields of an antenna. The proposed model requires only the knowledge of three parameters (Initial Current I_0, Damping coefficient α and the radial parameter τ) and hence considerably reduces the computational time and space as its results do not depend on the number of functions involved. The new technique is also developed to take account of the conductivity property of the surrounding medium. Hence this technique can be used in field prediction for antennas employed in medical imaging applications. Initial results obtained from the new technique show good correlation in comparison with the MOM technique.

1. Introduction

One of the fundamental problems in predicting the electromagnetic field generated by a semi rigid co-axial wire "monopole antenna" at a particular point in space has been addressed in this paper. This requires, in particular, the knowledge of the current density across the wire. Conventionally the Method of Moments (MOM) technique are used for this purpose whereby integral equations are discretised to predict the current across the wire[1, 2]. Although this technique is versatile it requires a large amount of computation *(N x N)* as the accuracy of the numerical result directly proportional to the number of basis functions N involved [3]. In addition, the MOM technique cannot be easily employed for applications involving media other than free space, such as medical imaging application, which require the use of coupling medium [4, 5]. Because of the conductive nature of the coupling medium the current prediction along the wire using the MOM technique becomes very laborious. More specifically $(N\,x\,N)^2$ computations are required.

In this paper a new technique has been proposed to model the current across a monopole antenna. The proposed model requires only the knowledge of three parameters for the prediction of the current along the wire; hence, reducing the computational time drastically as its results does not depend on the number of function involved. Also the new model will be useful to predict the antenna fields and impedance in different coupling media easily. Consequently this technique can be used in field prediction for antennas employed in medical imaging application.

This paper is organised as follows: The next section details the MOM technique procedures to calculate the current across the monopole antenna involving the Pocklington integral equations. Section 3 describes the new mathematical model. In Section 4 the results of the integral equation method is compared with the results obtained through our new mathematical model. Finally the conclusions on the model obtained are presented.

2. Method of Moment technique

In this section the so-called Method of Moments (MOM) technique is recalled. Consider an equation of the form:

$$Lf = g \tag{1}$$

where L is a continuous linear operator such as the integral operators or the differential operators, function f is the unknown to be determined and g represents a known excitation. If a unique solution exists, it is given by $f= L^{-1}g$, where $L^{-1}$is the inverse operator. L^{-1} cannot be determined analytically and thereby numerical solution is sorted [1]. MOM is one such numerical solution in which the continuous equation is discretised into matrix equations. The specific discretisation places a limit on the accuracy of a numerical result for a fixed number of basis functions and determines whether or not the numerical result will converge to the exact solution as the number of basis function increased. The solution to the above equation (1) can be given as

$$f = \sum_{n=1}^{N} \alpha_n \beta_n \tag{2}$$

Where the functions β_n are known basis functions defined on the domain of L and the scalars α_n are unknown coefficients to be determined. In what follows the MOM technique will be employed to calculate the current across the monopole.

2.1. Current along the wire using MOM technique

Consider a straight wire of length l and of radius a, the wire is placed along the z axis as shown in Figure 1.

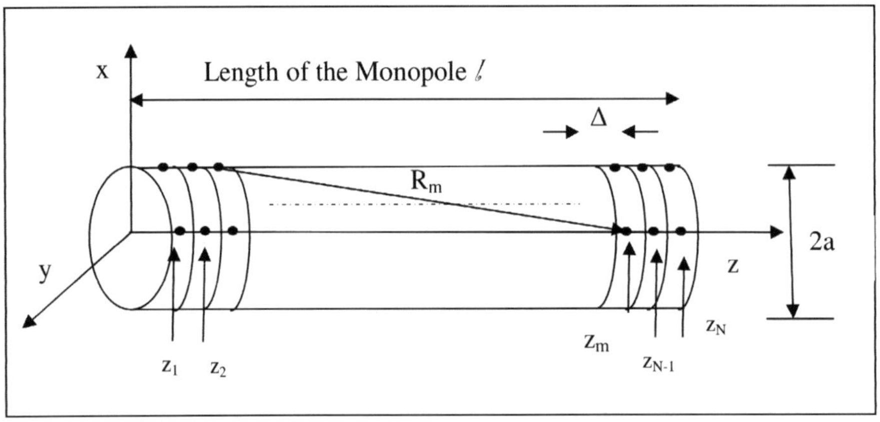

Figure 1: Segmented Wire.

A linear charge distribution $\rho(r')$ creates an electric potential $V(r)$ across the wire[6] which is given by

$$V(r) = \frac{1}{4\pi\varepsilon_0} \int \frac{\rho(r')}{R} dl' \tag{3}$$

Where $r(x', y', z')$ denotes the source coordinates, $r(x, y, z)$ denotes the observation coordinates dl' is the path of integration and the R is the distance from any point on the source to the observation point, which is given by

$$R(r - r') = |r - r'| = \sqrt{(x - x')^2 + (y - y')^2 + (z - z')^2} \tag{4}$$

The unknown charge distribution $\rho(z')$ can be approximated by an expansion of N known terms with constant, but unknown coefficients, that is,

$$\rho(z') = \sum_{n=1}^{N} \alpha_n g_n(z') \quad (5)$$

This expansion for the unknown charge distribution can be substituted in equation (3) and after interchanging the summation and integration the electric potential can be written as

$$4\pi\varepsilon_0 = \sum_{n=1}^{N} \alpha_n \int_0^l \frac{g_n(z')}{\sqrt{(z-z')^2 + a^2}} dz' \quad (6)$$

Where α_n are the scalars same as that of equation (2). The wire is now divided into N uniform segments, each of length $\Delta = l/N$ the $g_n(z')$ function is referred as basis functions and is given by

$$g_n(z') = \begin{cases} 0 & z' < (n-1)\Delta \\ 1 & (n-1)\Delta \leq z' \leq n\Delta \\ 0 & n\Delta < z' \end{cases} \quad (7)$$

Replacing z in equation (6) by a fixed point such as z_m on the surface of the wire and choosing N observation points $z_1 \ldots z_N$ on the surface of the wire each at the centre of each Δ length element to form N linearly independent equations to obtain the solutions for $a_1 \ldots a_N$ equation (6) can be expressed in matrix notation as

$$[I_n] = [Z_{mn}]^{-1}[V_m] \quad (8)$$

Where

$$[V_m] = \begin{pmatrix} 4\pi\varepsilon_0 \\ \vdots \\ 4\pi\varepsilon_0 \end{pmatrix}, [I_n] = \begin{pmatrix} a_1 \\ \vdots \\ a_1 \end{pmatrix}$$

And

$$Z_{mn} = \int_0^l \frac{g_n(z')}{\sqrt{(z_m - z')^2 + a^2}} dz' = \int_{(n-1)\Delta}^{n\Delta} \frac{1}{\sqrt{(z_m - z')^2 + a^2}} dz'$$

Equation (8) gives the current across the wire of length l and radius a using the MOM technique. In the next subsection the Pocklington integrodifferential equation will be employed to calculate the current of the semi-rigid cable.

2.2. Semi-rigid cable current using Pocklington Integrodifferential equation

In the case of the monopole antenna constructed by stripping the outer conductor of the semi rigid coaxial cable (Figure 2) and exposing the centre conductor, the wire exposed will experience an incident field across the surface caused by the coaxial aperture of the outer conductor.

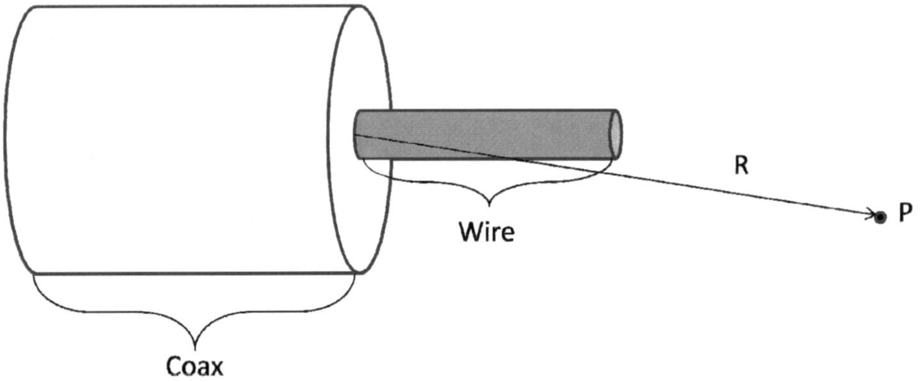

Figure 2: Semi Rigid Coax Monopole Antenna

Figure (3) shows the coaxial apertures and its equivalent magnetic frill given by equivalence principle which states *"two sources producing the same field within a region of space are said to be equivalent within that region"*. Thereby the coaxial aperture is replaced by an equivalent magnetic frill to calculate the fields from coaxial apertures. The magnetic frill consists of a circumferentially directed magnetic current density that exists over an annular aperture with inner radius a , which is usually chosen to be the radius of the wire, and an outer radius b. Figure 4 shows the segmented wire with the magnetic frill.

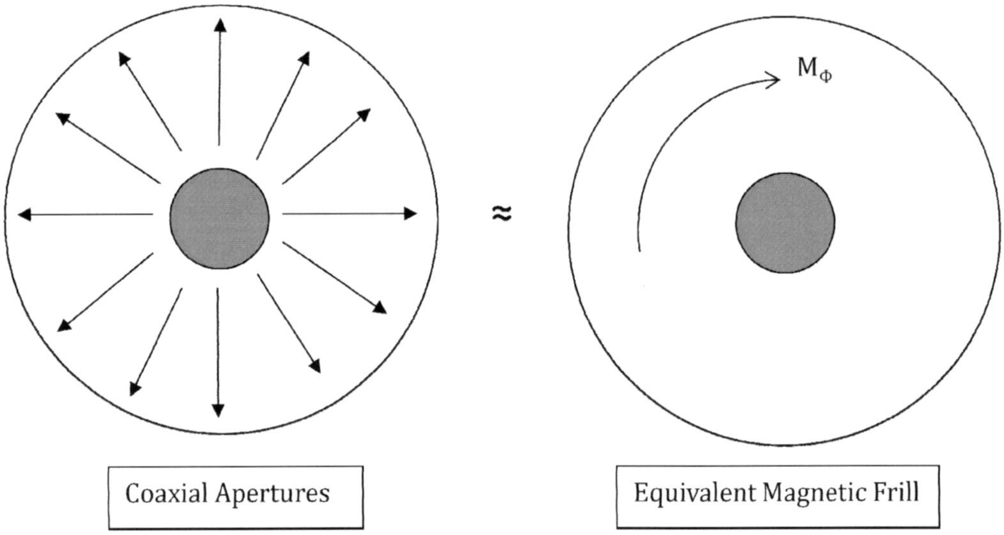

Figure 3 Equivalence Principle

The Pocklington integrodifferential equation[7] (11) can be used to determine the equivalent filamentary line source current of a thin wire, by knowing the incident field on the surface of the wire[8].

$$\int_0^l I_z(z')\frac{e^{-jkR}}{4\pi R^5}[(1+jkR)(2R^2-3a^2)+(kaR)^2]\,dz' = -j\omega\varepsilon E^i_{z'}(\rho=a) \qquad (11)$$

Where for observations along the centre of the wire *(ρ=0)*

$$R=\sqrt{a^2+(z-z')^2}$$

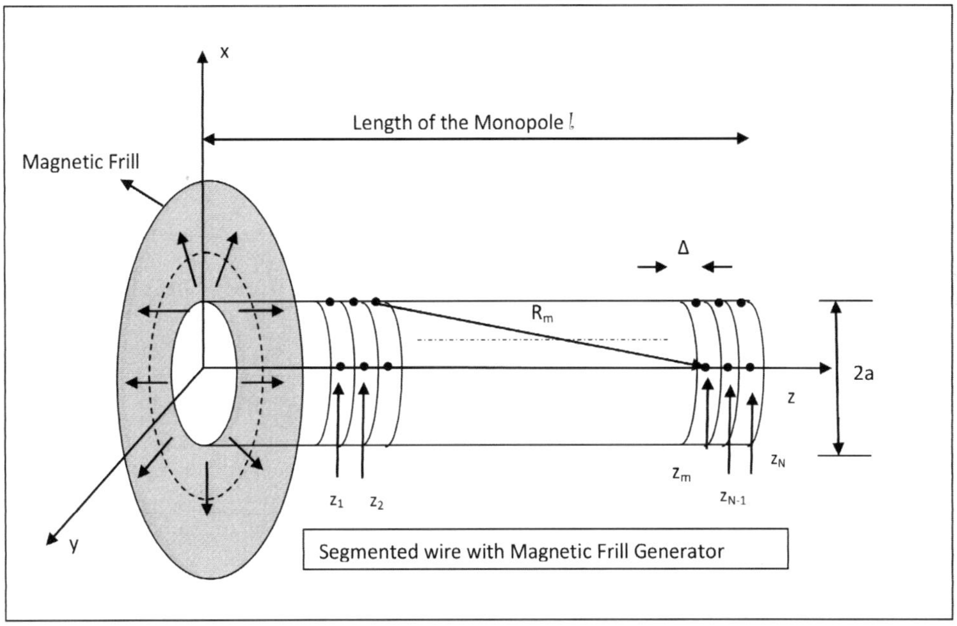

Figure 4: Segmented Wire with Magnetic Frill

Over the annular aperture of the magnetic frill generator, the electric field is represented by the TEM mode field distribution of a coaxial transmission line and is given by

$$E_f = \widehat{a_\rho}\frac{V_i}{\ln(b/a)} \quad a\le\rho'\le b$$

Therefore the corresponding equivalent magnetic current density M_F for the magnetic frill generator used to represent the aperture is equal to

$$M_F=-2\hat{n}\,x\,E_f=-2\hat{a}_z\,x\,\hat{a}_\rho E_\rho=-\hat{a}_\phi\frac{V_i}{\rho'\ln(b/a)} \quad a\le\rho'\le b \qquad (12)$$

The field generated by the magnetic frill generator of (12) on the surface of the wire along the axis *(ρ=0)* is given by [9]

$$E^i_z(\rho=0,0\le z\le l)=-\frac{V_i}{2\ln(b/a)}\left[\frac{e^{-jkR_1}}{R_1}-\frac{e^{-jkR_2}}{R_2}\right] \qquad (13)$$

where

$$R_1=\sqrt{z^2+a^2}\quad R_2=\sqrt{z^2+b^2}$$

Knowing the incident field, the current across the wire can be calculated using (13). Using the MOM technique the current $I_z(z')$ can be given by

$$\mathrm{I_z}(z') = \sum_{n=1}^{N} \alpha_n \mathrm{g_n}(z')$$

Replacing $I_z(z')$ in equation (11)

$$\int_0^1 \sum_{n=1}^{N} \alpha_n \mathrm{g_n}(z') \frac{e^{-jkR}}{4\pi R^5}[(1+jkR)(2R^2-3a^2)+(kaR)^2]\,dz' = -j\omega\varepsilon E_{z'}^{i} \quad (14)$$

After piecewise integration the current across the n segments is given by

$$[I_n] = [Z_{mn}]^{-1}[V_m] \quad (15)$$

where voltage at m^{th} segment can be written as

$$V_m = E_z^i \Delta$$

and Z_{mn} is given by

$$[Z_{mn}] = \int_{n\Delta}^{(n+1)\Delta} \frac{e^{-jkR(z_m,z'_n)}}{4\pi R^5(z_m,z'_n)} \Big\{\big(1+jkR(z_m,z'_n)\big)\big(2R(z_m,z'_n)^2-3a^2\big) + \big(kaR(z_m,z'_n)\big)^2\Big\}$$

Knowing the values for Z_{mn} and V_m across the segment, the current across the wire I_n can be calculated using equation (15). This equation calculates the current across the wire of length l and radius a using the MOM technique. However this technique is a time consuming process which requires NxN computations to predict the current. Also as the surrounding medium becomes conductive the calculation becomes complex thereby increasing the computational time to $(N\ x\ N)^2$. In the next section the mathematical model for the current distribution developed in the paper is presented.

3. The Proposed Mathematical Model

The extent of the limitation of the traditional MOM technique discussed above is directly proportional to the number of basis functions involved. In antenna field calculations this limitation is related to the segmentation of the wire. As the segment length Δ decreases, the number of segments N increases. As N increases, the accuracy of the current distribution along the wire increases, however this also increases the complex nature of the matrix equation involved, and thereby increases the computational time. This limitation of the MOM technique increases as we address the issue of field calculation of antenna in conducting media. In a conducting medium the dielectric property of the medium is complex[10] and it takes into account the loss tangent of the medium. Thereby the propagation constant k will be of complex nature,

$$k = k' + jk''$$

and this will in turn increase the computational time from NxN to $(NxN)^2$. In order to address the limitations of the MOM technique for current distribution calculation, a new mathematical model is proposed. The new model aims to reduce the computational time and the tedious nature of the MOM equations explained in the previous section. The expression for the new model is given below in equation 16.

$$I(z) = I_0 e^{-\alpha z}\sin(k(l-z)) + f(z,\tau) \quad (16)$$

Equation 16 consists of two parts; the first part,

$$I_0 e^{-\alpha z}\sin(k(l-z))$$

accounts for the damping in the current distribution curve of Figure 5. This characterises the effect of the surrounding medium of the wire. The current distribution curve in Figure 5 is of the wire of length λ/2 in free space. In this case the damping coefficient α is zero and its value changes as the surrounding medium changes. This is very effective for applications involving coupling medium with complex dielectric properties, such as medical imaging applications. This part also provides the overall shape of the current distribution curve in Figure 5. This part of the equation is similar to that of the current distribution expression given in [1]. The final part of the expression is generally given by,

$$f(z,\tau) = \begin{cases} d_0 + \frac{\tau}{4}\sin\big(2k(l-z)\big), & for\ l = (2w+1)\ \frac{\lambda}{4} \\ d_0 + 2\tau\sin\big(2k(l-z)\big), & for\ l = 2w(\lambda/4) \end{cases}$$

where d_0 is the dc component and w is a positive integer. This part accounts for the variation due to the radii of the wire; it acts like the dc term in the expression. It also provides the delay element in the current distribution curve in Figure 5.

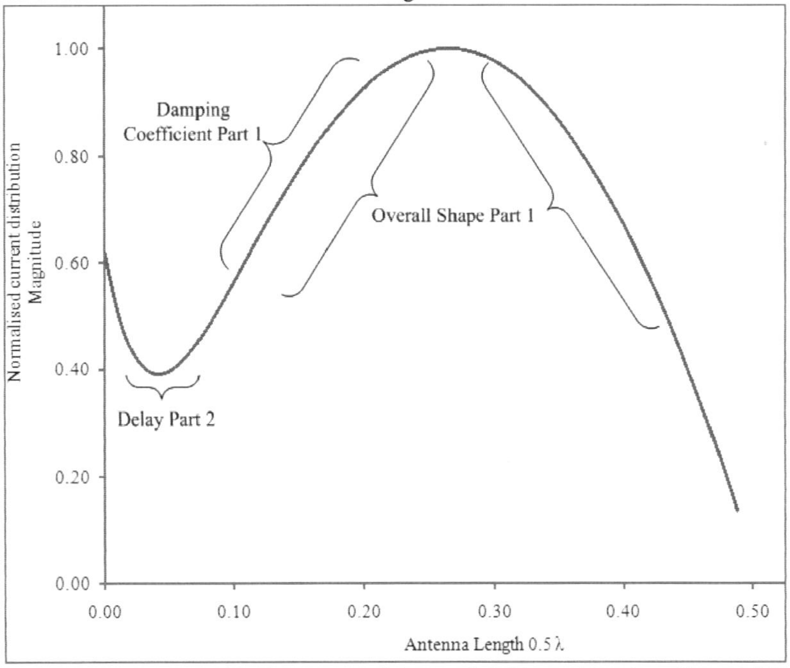

Figure 5: current distribution curve of the semi rigid coaxial wire of length λ/2.

This new mathematical model decreases the computational time as it depends on only three parameters; Initial current I_0, damping coefficient α and radial parameter τ. Initial current I_0 is the current at the first segment of the wire, damping coefficient α characterises the conductivity of the surrounding medium. It is this parameter of the expression which makes this model suitable for the predicting the current distribution of the wire in different surrounding media other than free space. And finally, τ is a parameter related to the radius of the wire.

4. Initial Results and Model Validation

The ability to predict the current across the monopole antenna using this new mathematical model has been analysed by predicting the current distribution of two different lengths of wire. Length of quarter wavelength (λ/4) and three quarter wavelength (3λ/4) monopoles are considered. The predicted current distributions of the above lengths of wire using conventional technique were available to compare with the results from the new mathematical model. Antenna radius a is given as 0.005λ and the outer radius b is related to the inner radius by $b = 2.3a$, and the wire is considered along the z axis. For the conventional method the number of basis function is kept at 80, i.e. the wire along the z axis is discretised to 80 segments. The initial current parameter I_0 of the new model is measured for different lengths of the antenna. Using these initial current values the other two parameters, damping coefficient α and radial parameter τ, are calculated using optimisation techniques.

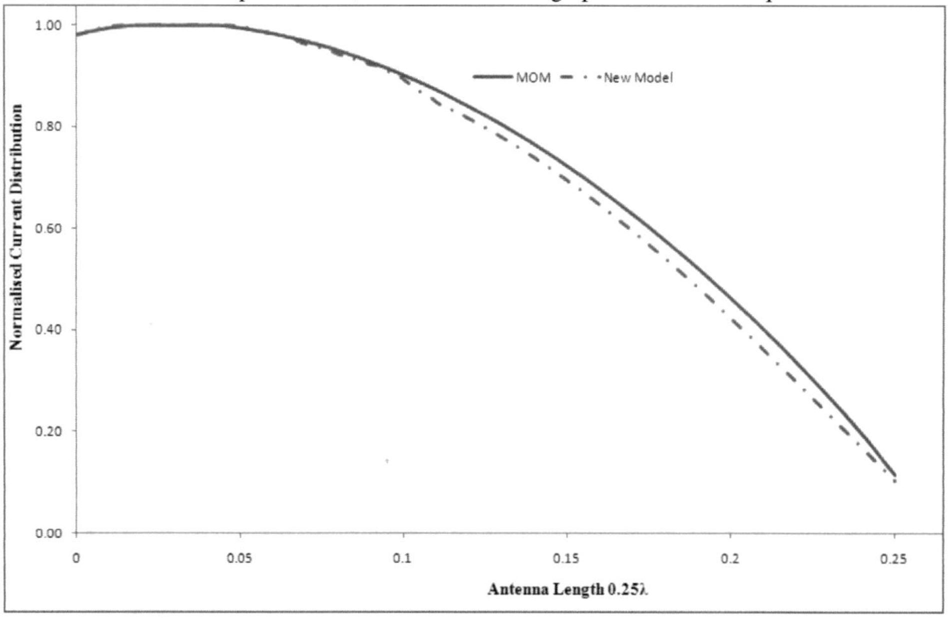

Figure 6: current distribution curve of the semi rigid coaxial wire of length λ/4.

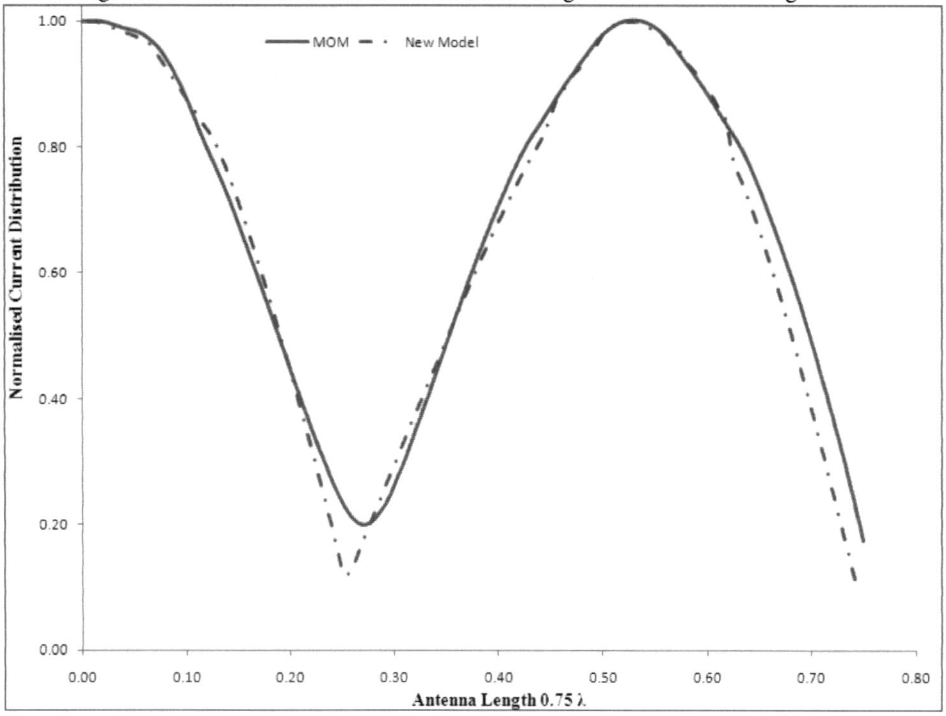

Figure 7: current distribution curve of the semi rigid coaxial wire of length 3λ/4.

Figures 6 & 7 shows the result of both the MOM technique and the new model. As it is shown there is a good correlation between the two techniques for the two different lengths of the wire. From the current values across the wire, the vector potential A at a distance P (refer to Figure 2) can be calculated using the expression

$$A(x,y,z)=\frac{\mu}{4\pi}\int_c I_c(x',y',z')\frac{e^{-jkR}}{R}dl'$$

Once the vector potential is known the electric field E at the point P due to the wire can be calculated. Thus the electric field of a monopole antenna can be calculated using our new model, because of the less complicated nature of the new model and its flexibility to predict the current distribution in media other than free space (damping coefficient α), this can be extended to surroundings involving coupling medium.

5. Conclusion

This paper presents a new model to predict the current across the monopole antenna thereby being useful in predicting the electromagnetic field at a particular point in space. The new model is computationally less tedious and laborious than the traditional MOM method. The initial results have shown a good correlation between the two models and thereby the new model can be used in monopole field prediction in coupling medium environment. Further work will investigate the validity of the new model in media such as saline solution, vegetable oil and other coupling liquids used for medical applications. It will also refine the $f(z,\tau)$ term.

References

1. Balanis, C.A., *Antenna Theory: Analysis and Design Second Edition.* 1997, New York: John Wiley&Sons, Inc.
2. Lilly, J.D., *Application of the Moment Method to Antenna Analysis*, in *Department of Electrical Engineering*. 1980, West Virginia University.
3. Peterson, A.F., S.L. Ray, and R. Mittra, *Computational Methods for Electromagnetics*. 1998, Oxford: Oxford University Press.
4. Smith, D., et al., *Medical Imaging using a Microwave Indirect Holographic Technique.* Mediterranean Microwave Symposium (MMS), Genova, Italy, 2006(w).
5. Elsdon, M., et al., *Microwave Holographic Imaging of Breast Cancer.* IEEE International Symposium on Microwave, Antenna, Propagation and EMC Technologies for Wireless Communications, 2007.
6. Cheng, D.K., *Field and Wave Electromagnetics.* 1989, Reading, MA: Addison-Wesley. 97.
7. Burgess, R.E., *Aerial Characteristics.* Wireless Engineering, 1944. 21: p. 154-160.
8. Thiele, G.A., *Wire Antennas*, in *Computer Techniques for Electromagnetics*, M. R, Editor. 1973, Pergamon: NewYork. p. Chapter 2 p:7-70.
9. Tsai, L.L., *A Numerical solution for the Near and Far Field of an Annular Ring of Magnetic Current.* IEEE Transactions on Antennas and Propagation, 1972. AP-20(5): p. 569-576.
10. Foti, S.J., et al., *A Water-Immersed Microwave Phased Array System for Interrogation of Biological Targets.* Medical Applications of Microwave Imaging, ed. L.E. Larsen and J.H. Jacobi. 1986, New York: IEEE Press. 148-166.

Journal of Physics: Conference Series **255** (2010) 012002 doi:10.1088/1742-6596/255/1/012002

An investigation into the use of a mixture model for simulating the electrical properties of soil with varying effective saturation levels for sub-soil imaging using ECT

R R Hayes[1], P A Newill[1], F J W Podd[1], T A York[1], B D Grieve[1], O Dorn[2]

[1]SSUIC, School of Electrical and Electronic Engineering, University of Manchester, UK

[2]Inverse Problems Group, School of Mathematics, University of Manchester, UK

Email: Robert.Hayes@postgrad.manchester.ac.uk

Abstract. A new visualisation tool is being developed for seed breeders, providing on-line data for each individual plant in a screening programme. It will be used to indicate how efficiently each plant utilises the water and nutrients available in the surrounding soil. This will facilitate early detection of desirable genetic traits with the aim of increased efficiency in identification and delivery of tomorrow's drought tolerant food crops. Visualisation takes the form of Electrical Capacitance Tomography (ECT), a non-destructive and non-intrusive imaging technique. Measurements are to be obtained for an individual plant thus allowing water and nutrient absorption levels for an individual specimen to be inferred. This paper presents the inverse problem, discusses the inherent challenges and presents the early experimental results. Two mixture models are evaluated for the prediction of electrical capacitance measurement data for varying effective soil saturation levels using a finite element model implemented in COMSOL Multiphysics. These early studies have given the research team an understanding of the technical challenges that must now be addressed to take the current research into the world of agri-science and food supply.

1. Introduction

Crops are one of the world's major water consumers [1], as climate change leads to water scarcity there is an increase in competition for water. This severely limits irrigation and constrains food production. In the face of climate change the ability to rapidly identify new plant varieties that will be tolerant to drought, and other stresses, is going to be key to breeding the food crops of tomorrow. With large areas of Africa and South America expecting major water scarcity by 2025 [2], it is very important to develop crops that can survive in these conditions. Currently, above soil features (phenotypes) are monitored in industrial greenhouses and field trials during seed breeding programmes so as to provide an indication of which plants have the most likely preferential genetics to thrive in the future global environments [3]. These indicators of "plant vigour" are often based on loosely related features which may be straightforward to examine, such as an additional ear of corn on a maize plant, but which are labour intensive and often lacking in direct linkage to the required crop features.

A new visualisation tool is being developed for seed breeders, providing on-line data for each individual plant in a screening programme indicating how efficiently each plant utilises the water and nutrients available in the surrounding soil. It will be used as an in-field tool for early detection of desirable genetic traits with the aim of increased efficiency in identification and delivery of tomorrow's drought tolerant food crops.

Imaging the moisture distribution in the sub-soil environment carries a number of challenges. For in-field measurements any instrumentation must be portable, relatively low cost, internally powered and weather resistant. High resolution imaging modalities such as magnetic resonance imaging (MRI) and X-ray computed tomography (CT), while offering superior image quality, cannot be used in this capacity as the technology is costly (>£30k) and requires large high powered instrumentation (>500W). Therefore a compromise to reduce size and cost at the expense of image quality must be made. For this reason, electrical impedance imaging is an ideal candidate technology.

This modality however poses its own significant technical challenges in light of the intricacy of the medium. Electrical impedance imaging is a soft field reconstruction problem. That is to say that the regions of interaction are altered by the sample under investigation. It also suffers from being an ill-posed problem. Furthermore the medium (soil) exhibits both conductive (wet) dominating phases and dielectric (dry) dominating phases. The implication of this is the formation and breaking of conducting paths between measurement electrodes and a transition between the dominance of the real and imaginary impedance components. As a result of water loss, soil is prone to shrinking and cracking which further complicates the measurement process due to the quality of the connection at the contact points between the medium and the electrodes.

The forward solution for simulating water movement in soil measured via electrical capacitance tomography (ECT) is a complex problem and requires the consideration of several theoretical backgrounds. For this reason simulation is a valuable tool. Simulation of plant growth, root water uptake and soil water transport and the integration of these models has been an area of high activity over recent years. A number of realistic root growth models have been developed and implemented in MATLAB (including meshing functions) [4][5] and ANSI C, respectively [6]. The R-SWMS code implemented in FORTRAN 90 integrates root growth, root water uptake and soil water transport models [7]. In addition to this several commercial products such as HYDRUS 3D [8] and COMSOL Multiphysics [9] allow the simulation of soil water transport in three dimensions.

As a first step, separate ECT and hydrology models are used. However in the future these will be combined into a multiphysics model to enable joint reconstructions. Multiphysics simulation allows the relationships between different physics backgrounds to be coupled. Hence providing a means by which the required models can be combined in one software package, this is advantageous as it negates the need for interfacing two separate software packages and potentially reduces computation time. Simulation also facilitates the optimisation of sensor design by allowing a number of sensor geometries to be compared, while avoiding error introduced by measurement instrumentation as the signal-to-noise levels and power density functions can be set precisely. In addition, simulation facilitates the comparison and optimisation of image reconstruction algorithms as the 'unknown' permittivity distribution that is to be reconstructed can be precisely defined. Furthermore, the desire is to incorporate the simulation model into the reconstruction process so that plant water uptake can be directly estimated from the time varying moisture distribution.

COMSOL Multiphysics has been used to simulate capacitance data for a soil fluid flow experiment and will further facilitate the implementation of a multiphysics based forward solver and potentially the use of prior information in the inverse problem.

2. Background

Simulation of the electrical capacitance measurements in this application is a multiphysics problem combining, flow through a porous media to describe the movement of water through a soil column, and electrostatics for the calculation of the electrical fields resulting from the moisture distribution.

Journal of Physics: Conference Series **255** (2010) 012002 doi:10.1088/1742-6596/255/1/012002

2.1. Flow in porous media

The Richards equation governs the saturated-unsaturated flow of water in non-swelling soils [10] and is given by

$$(C+S_e S)\frac{\partial H_p}{\partial t}+\nabla\cdot\left(-K\nabla\left(H_p+D\right)\right)=Q_s \tag{1}$$

where C is specific moisture capacity (m^{-1}), S_e is the effective saturation of the soil, S is a storage coefficient (m^{-1}), H_p is the pressure head (m), D is the vertical elevation (m), t is time (d), K is the hydraulic conductivity (m/d) and Q_s is a fluid source defined by volumetric flow rate per unit volume of soil (d^{-1}).

Applying the Richards equation for a point injection with a given pressure head to a rectangular soil geometry allows the calculation of the soil water distribution at a given time. Figure 1 is an example soil water distribution obtained from the solution of the Richards equation for a rectangular soil geometry simulated using the COMSOL Multiphysics Earth Science module. The soil retention characteristics and saturated hydraulic conductivity in this example were chosen arbitrarily.

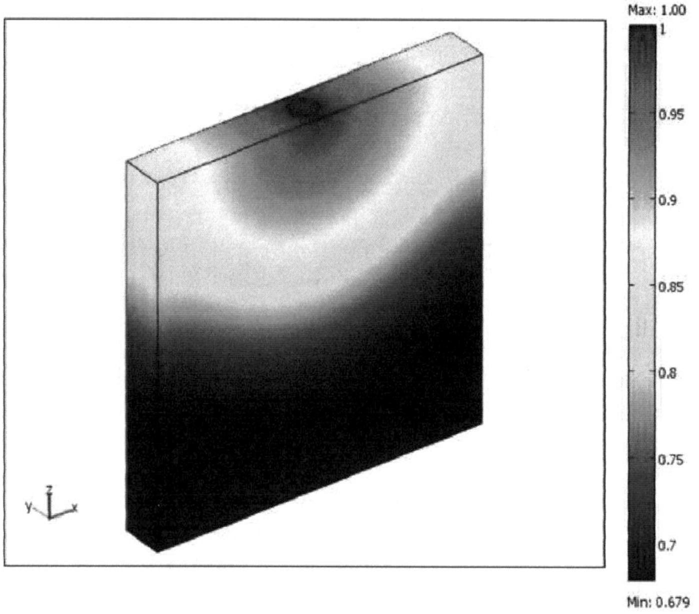

Figure 1. 3D Simulated soil water distribution at a time, t, for a point injection.

Root water uptake can be included in the Richards equation via a sink term. If a root network is considered as a series of nodes and interconnecting segments, the Doussan model can be used to calculate the water potential at each of the root nodes from a system of equations expressed in terms of the length of the interconnecting segments, the Xylem conductivity, the water potential of the surrounding soil, the radial conductivity and the surface area of the interconnecting segments [11].

2.2. Electrostatics

The forward problem in ECT is to calculate the potential distribution for a known dielectric constant and then to determine the corresponding capacitances measured at the surface. Assuming no free charge conditions, the electric scalar potential, V, between the electrodes mounted on the vessel satisfy Poisson's equation [12]

$$-\nabla \cdot \left(\varepsilon_0 \varepsilon_r \nabla V\right) = \rho \tag{2}$$

where ε_0 and ε_r are the permittivity of free space and the dielectric constant respectively, ρ is the space charge density. The electric field (E) and displacement field (D) can be obtained from the potential gradient

$$E = -\nabla V \tag{3}$$

$$D = \varepsilon_0 \varepsilon_r E \tag{4}$$

For the plastic surfaces of the vessel, the conditions of zero surface charge are applied at the boundary, thus

$$n \cdot D = 0 \tag{5}$$

A finite element based implementation of this mathematical model was validated experimentally (see section 3.1).

2.3. Multiphysics Coupling

The dielectric permittivity of a partially saturated soil matrix varies as a function of water content. The implication of being able to describe this relationship mathematically is the ability to infer the water content in a soil sample via electrical measurement.

Electromagnetic techniques for inferring soil moisture content, such as time domain reflectometry (TDR) and capacitance methods [13] rely on this strong dependence of electrical signals on dielectric permittivity [14]. The permittivity can be related to soil water content using a calibration curve [15].

This consists of establishing the relationship between the signal provided by the instrumentation and the soil moisture content by using a reference method. This relationship is comprised of two subsequent relationships. Firstly the relationship between the probes output signal and the soil dielectric permittivity, which may be determined experimentally. Secondly the relationship between soil dielectric permittivity and soil water content. This may be determined using theoretical models or empirically [16].

One further interesting property of being able to mathematically model the relationship between soil moisture content and dielectric content is the potential to estimate electrical measurements (i.e. capacitance measurements) for a given soil sample based on a simulated soil moisture distribution. This provides a basis for the implementation of a multi-physics simulation.

Fluid flow and electrostatics models can be coupled via the dielectric constant, if the dielectric constant of soil can be predicted based on effective soil saturation. The dielectric constant vs. soil saturation curve may be obtained experimentally or predicted using theoretical methods.

A number of methods have been proposed to predict the dielectric constant of water saturated rocks and soils at varied soil saturation levels. Two of the more popular theoretical methods are the Topp model [17] and the Complex Refractive Index Method (CRIM) [18]. These methods have been tested extensively [19][20] and are commonly used in the calibration of dielectric soil moisture probes [14].

The popular empirical model known as the Topp model was developed by Topp et al by compiling data for many soils under varying moisture conditions [21]. This model is given by the third-order polynomial equation

$$\varepsilon = 3.03 + 9.3\theta + 146\theta^2 - 76.7\theta^3 \quad (6)$$

where ε denotes the effective dielectric constant of the soil mixture and θ denotes the water content. Extensive testing has shown the Topp model to be reasonably accurate for many soils [22]. However its validity has not been demonstrated over a full range of possible water contents and porosities [23].

The complex refractive index method (CRIM) involves using simple mixing laws to calculate the dielectric constant of a material (ε_T) given the dielectric constant (ε_i) and the volume fraction (V_i) as given by

$$\sqrt{\varepsilon_T} = \sum_i V_i \sqrt{\varepsilon_i} \quad (7)$$

This can be expanded for the case of a soil sample at a given level of water saturation (θ), where θ is the volume fraction of the pore space that is filled with water, the remainder being filled with air [24]. The system is composed of soil (ε_S), water (ε_W) and air (ε_A). Porosity is given by ϕ and the dielectric constant for a given soil saturation level is expressed as

$$\sqrt{\varepsilon_T} = (1-\phi)\sqrt{\varepsilon_s} + \theta\phi\sqrt{\varepsilon_W} + (1-\theta)\phi\sqrt{\varepsilon_A} \quad (8)$$

Assuming that the soil and air components have negligible conductivity, the dielectric constants of both soil and air can be assumed to be entirely real[24][25].

An obvious limitation of both models is that they assume localised homogeneity and complete mixing. They also do not take into account electro-chemical interactions amongst the constituent components [24].

This paper will discuss the validity of the FEM model and compare simulated capacitance data, based on the dielectric permittivity predicted by the Topp and CRIM models, with measured capacitance data (see section 3.2).

3. Experimental Procedure and Results

3.1. Validation of the FEM through observing the effect of water fill level

In order to validate an electrostatic simulation implemented in COMSOL Multiphysics, the capacitance of a rectangular vessel as described in figure 2 was measured for varying water fill levels. The vessel walls were manufactured to include two parallel plate electrodes of dimension 130mm x 85mm as shown in figure 3. These electrodes were insulated from the medium using a solder mask layer of 75μm thickness. Measurements were carried out using a Hewlett Packard 4192A Impedance analyser.

The dielectric constant of the solder mask layer was set to 3.56 (as defined by the manufacturer) in the FEM. Simulations were then carried out for fill levels ranging from 0mm (100% air) to 120mm (100% water) in 10mm steps. The dielectric constant of air was taken as 1.00059 [26] and the dielectric constant of water was taken as 79 at 20 degrees Celsius [27][28].

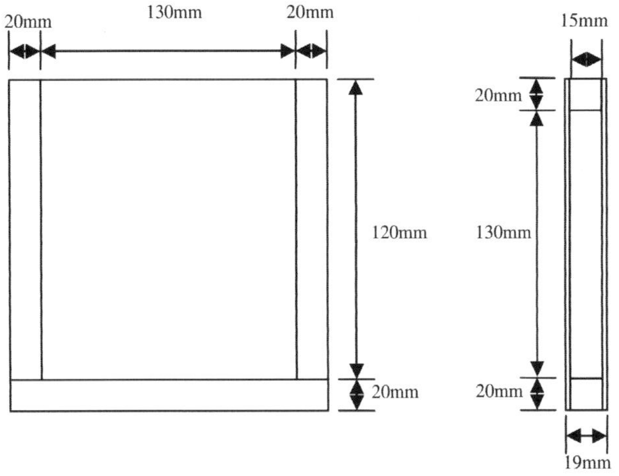

Figure 2. Schematic diagram of the rectangular vessel. Left: Front view, Right: Top view.

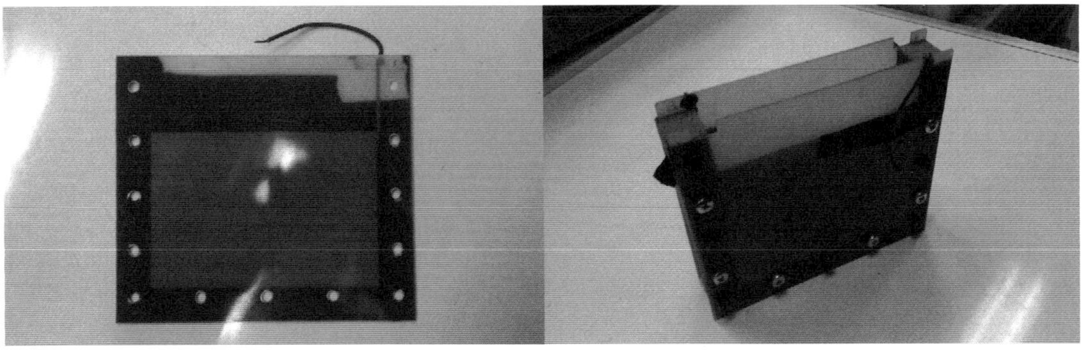

Figure 3. Left: Vessel wall showing electrode Right: Constructed measurement vessel

A comparison between measured and simulated data is shown in figure 4. The point at which both measurement sets cease to increase linearly is the point at which the water fill level exceeds the height of the electrode. Beyond this level, the change is in the fringing field only and variations are of ~11pF corresponding to approximately 3% of the total measured capacitance. The maximum error between simulation and measurement is 43.64% corresponding to an error of 5.26 pF. This error occurs when the vessel is empty, i.e. filled with air. This point is the lowest measured capacitance, hence the most sensitive to measurement error. The average error is 8.88%. Possible sources of error include experimental error and the absence of temperature data in order to account for the temperature dependency of the dielectric constant of water. Analysis of the potential sources of error is presented in section 3.3.

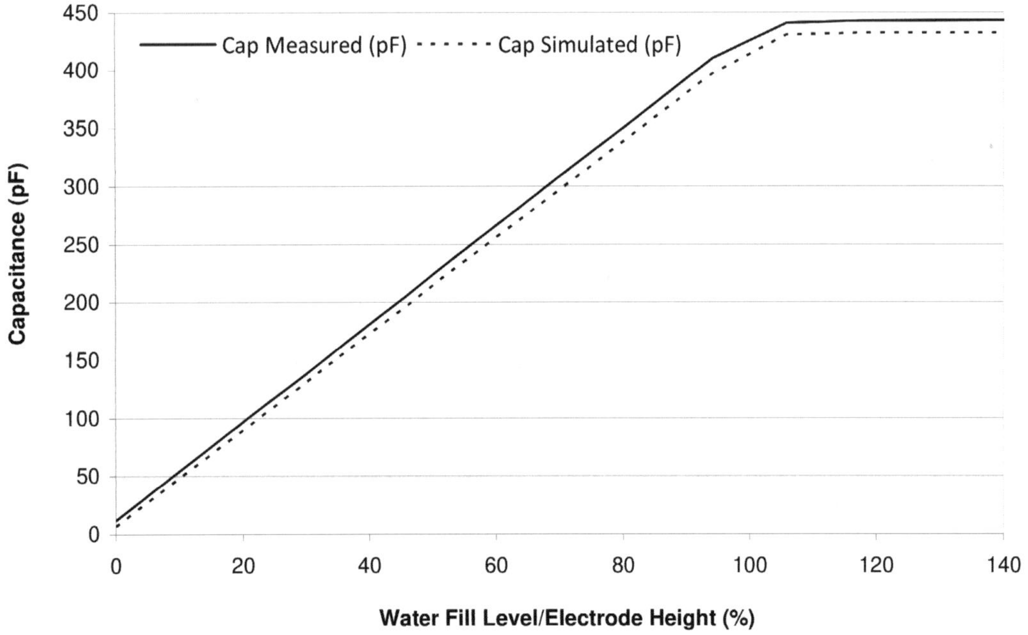

Figure 4. Plot of capacitance vs. fill level for measured and simulated capacitance data.

Figure 5 shows the simulated electric fields for a half filled vessel. Note the electric field is highest in the region of higher dielectric constant (i.e. the water) as represented by the largest arrows. The smaller arrows represent the weaker electric field in the air filled region.

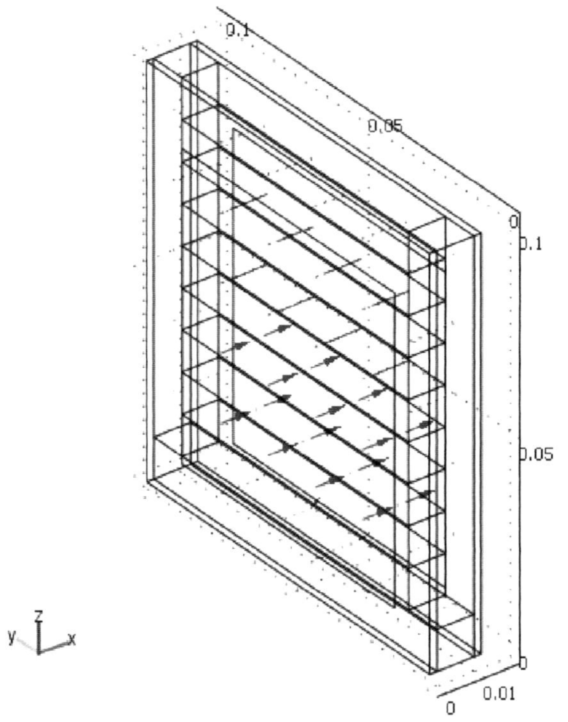

Figure 5. Simulated electric fields for a half filled rectangular vessel.

3.2. Varying soil saturation

To determine the validity of the Topp and CRIM models, the HP impedance analyser was also used to measure the capacitance of the vessel when filled with soil of varying levels of saturation. The dielectric constant of the soil matrix was predicted for the corresponding saturation levels using both the Topp and CRIM models. For the CRIM model porosity was taken as 0.54. This value was calculated from data provided by a soil analysis service offered by Forest Research Surrey using the following

$$\phi = 1 - \frac{\rho_{bulk}}{\rho_{particle}} \tag{9}$$

where ϕ is the porosity, ρ_{bulk} is the bulk density, defined as the average density of dry soil. $\rho_{particle}$ is the particle density, defined as the density of the particles that the soil consists of. It is equal to the dry density of the solid material comprising the soil matrix i.e. the mass of air is negligible [29]. The particle density of a soil may be estimated using the weighted average of the solid components [29] and is given by

$$\rho_{particle} = \rho_m X_m + \rho_{om} X_{om} \tag{10}$$

where $\rho_{particle}$ is total soil density, ρ_m and ρ_{om} are the mineral fraction density and organic matter fraction density respectively, X_m and X_{om} are the mineral volume fraction and the organic matter volume fraction respectively. The porosity value of 0.54 is consistent with values for silty soils [30].

The predicted dielectric constants from the Topp and CRIM models were substituted into the FEM and the respective capacitance measurements were calculated by simulation.

As the soil water content was increased, the capacitance measurements had two distinct regions of linearity, as can be seen in figure 6. Any non-linearity is likely to be caused by inconsistencies in the level of compaction of soil and/or any bowing of the vessel wall between subsequent measurements.

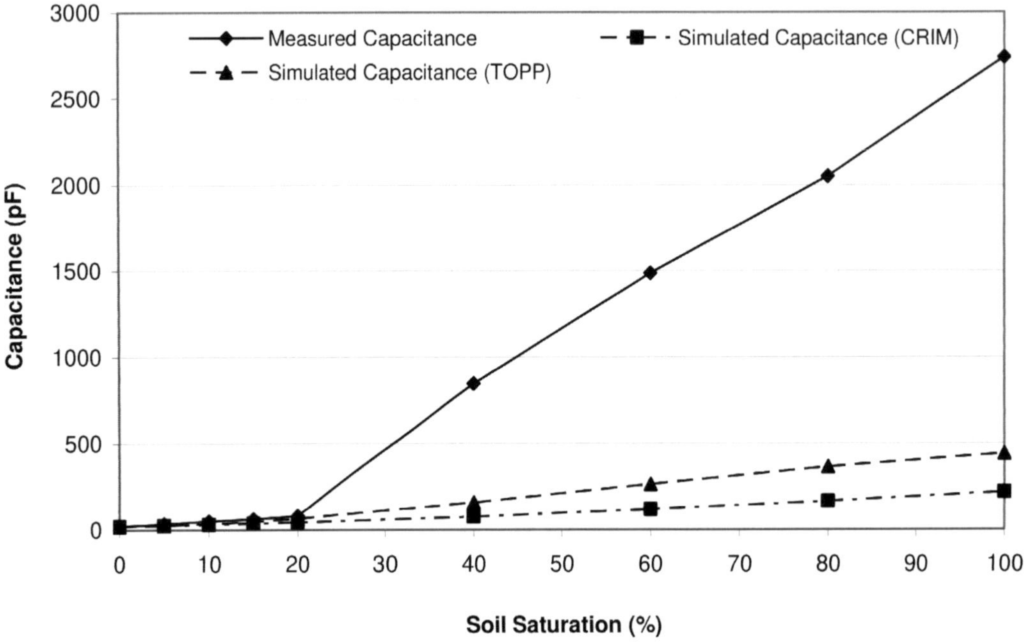

Figure 6. Measured and simulated capacitance measurements vs. soil saturation level.

Beyond a saturation level of ~20% there is a sudden increase in the slope. This discrepancy to the mathematical model is believed to be the result of resistive coupling on the reading from the HP impedance analyser for conductivities greater than 0.48 mS. Distilled water was used. However the mineral content of the soil increased the conductivity of the water from 0 mS to 0.75 mS. This hypothesis was investigated by repeating the procedure using a non-conductive liquid (Rapeseed oil) to saturate the soil. It is assumed that the ions present in the soil are not oil soluble and do not increase the conductivity of the oil. An equivalent experiment using "washed" sand and distilled water was also carried out.

The CRIM equation was used to predict the dielectric constant of the soil/sand matrix for varying saturation levels and substituted into the FEM to simulate the expected capacitance measurement data. This simulated data was compared with measured data. In this case, the dielectric constant of rapeseed oil was taken as 3.1 [31]. The Topp model was not considered in these experiments as it is formulated for a soil-water matrix and cannot be modified to describe other components.

As can be seen in figures 7 and 8, the change of slope is no longer present. This suggests that the hypothesis of instrumentation limitations for high conductivities appears valid. Any non-linearity in the measured data is likely due to variations in the level of compaction between subsequent measurements. Predictions based on the CRIM model had an average error of 3.77% and 5.25% for the soil-oil and sand-water experiments respectively as compared to measured data.

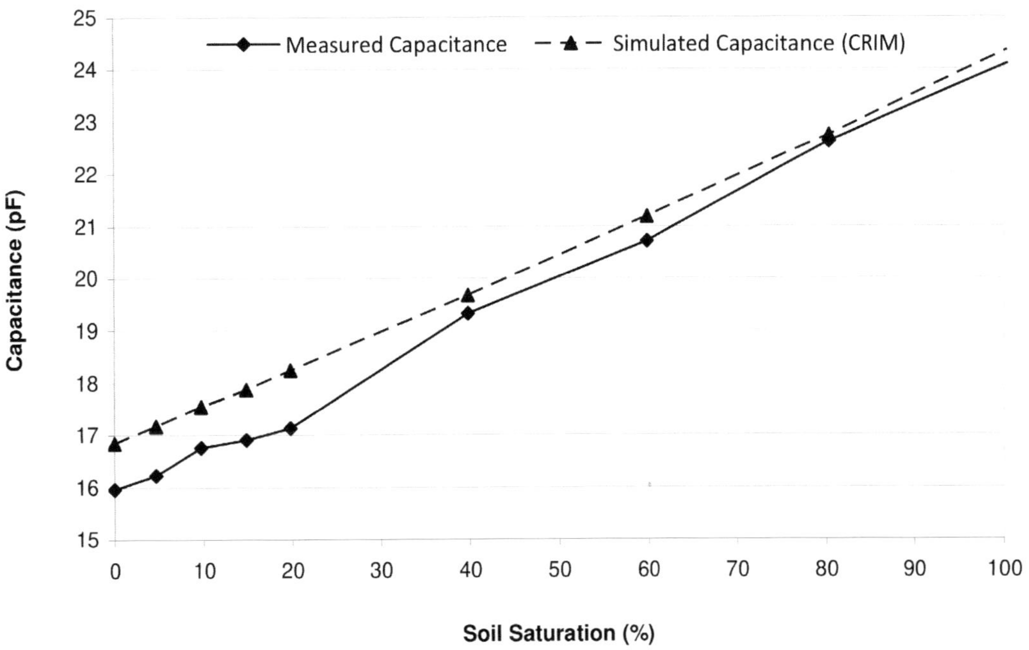

Figure 7. Measured and simulated capacitance measurements vs. level of soil saturation using rapeseed oil.

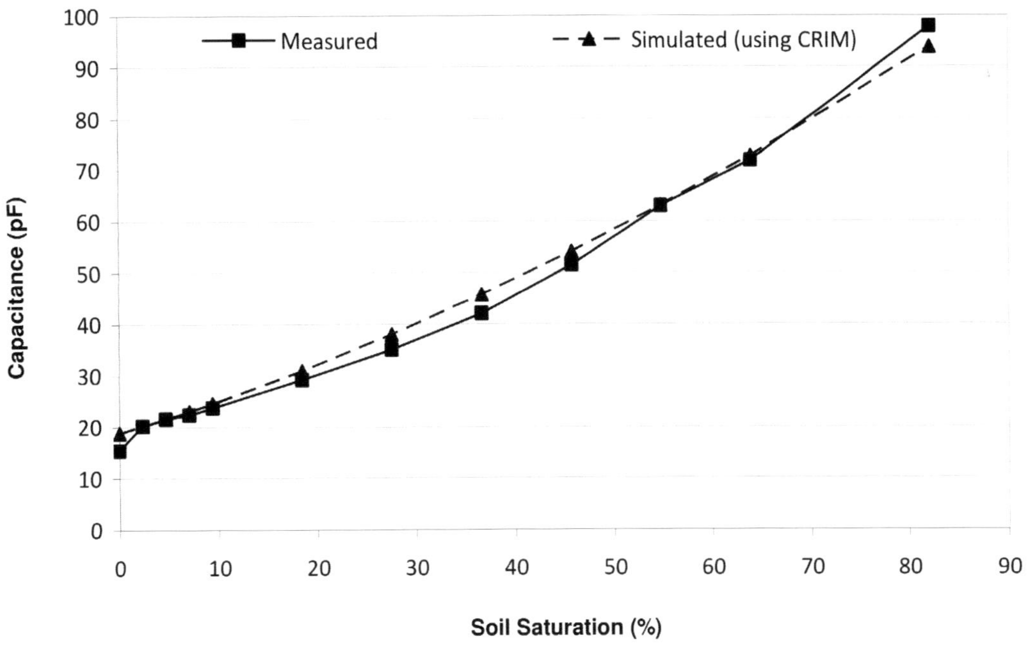

Figure 8. Measured and simulated capacitance measurements vs. level of sand saturation using distilled water

It is thought that at lower measurement frequencies the effects of soil type and soil conductivity are prevalent. The result is a requirement for higher excitation frequencies this is due to decreased conductive losses at higher frequencies hence reducing the effect of soil conductivity [32]. At present the measurement instrument is limited to 1MHz. The result of using a higher frequency is a simplification of the calibration process. In many cases a single calibration curve may be used for numerous soil types [33]. The literature suggests that using suitable calibration curves capacitance measurements can provide reliable results over the full typical range of saturation levels (0-60%) [13][34]. Typically soil dielectric measurement instruments operate in the 10-200 MHz region [13].

3.3. Investigating the source of error

In order to determine the sensitivity of measurement data with respect to temperature, the variability of the dielectric constant of water vs. temperature was investigated. The variation of dielectric constant of water with respect to temperature was modelled using the polynomial equation [35]

$$\varepsilon_w = 5321T^{-1} + 233.76 - 0.9297T + 0.1417\times10^{-2}T^2 - 0.8292\times10^{-6}T^3 \quad (11)$$

where ε_w is the dielectric constant of water and T is the temperature in Kelvin.

The modelled dielectric constant of water for the temperature range of 0 to 100 degrees Celsius is shown in figure 9. The standard parallel plate capacitor equation was then used to calculate the capacitance for the vessel fully filled with distilled water. The thickness and dielectric constant of the dielectric layer is known. The dielectric constant of water was taken as 79 at 20 degrees Celsius [35]. These values were taken as a reference. To determine the extent to which the measurements were sensitive to temperature changes of water, the dielectric constant of water was varied until the resulting expected capacitance changed by ~10% of the reference. This corresponds to a change of 12.6% in the value of the dielectric constant of water which is equivalent to a temperature rise of approximately 25-30 degrees Celsius. This results in a temperature of approx 50 degree Celsius. It is unlikely that such an error in temperature measurement occurred. Figure 10 shows the change in capacitance calculated for the temperature varied dielectric constant of water.

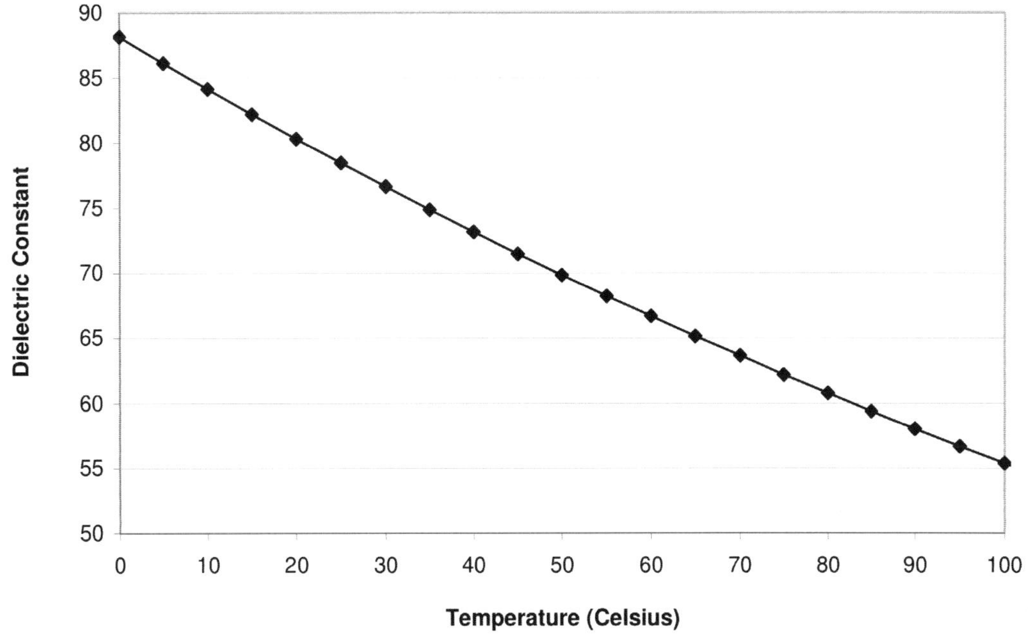

Figure 9. Variation of the dielectric constant of water with respect to temperature.

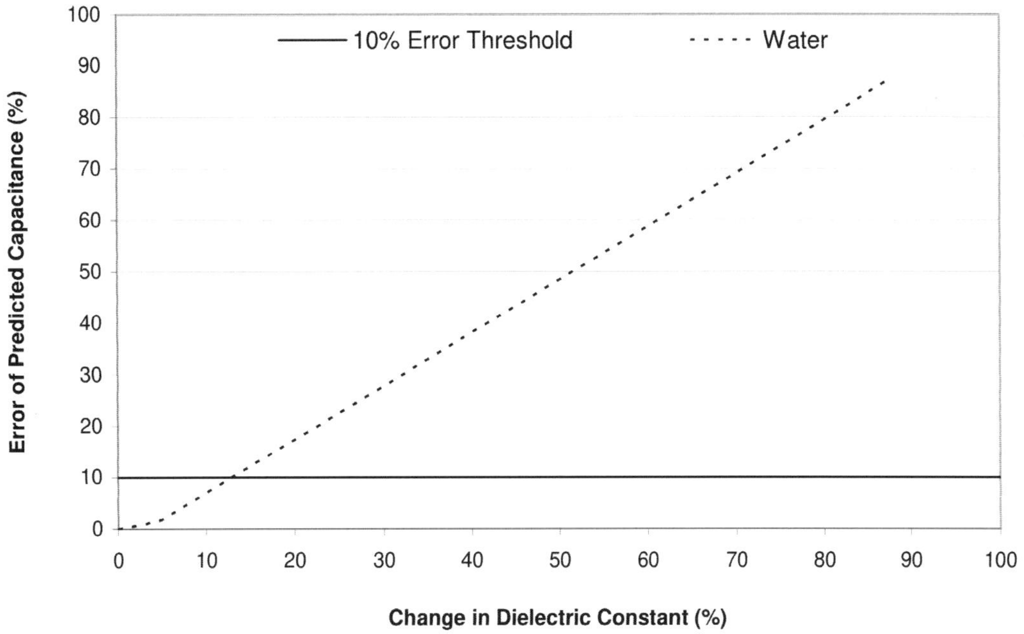

Figure 10. Error in predicted capacitance vs. temperature induced drift in dielectric constant of water for determining the extent to which the temperature dependence of the dielectric constant of water contributes to measurement error.

4. Conclusions

4.1. Initial findings

Experiments have shown that the simulation is able to predict changes in capacitance due to varying levels of soil saturation based on two methods for predicting the dielectric constant of varying soil water saturation levels.

The electrostatic model has been shown to exhibit an average error of 8.88% for varying water fill levels in comparison to measured data. Predictions generated using both the Topp and CRIM models were compared to measured data for varying soil saturation levels. It has been noted that beyond 20% soil saturation, there are difficulties in measuring the capacitance of the wetted soil. This has been attributed to the limitations of the measurement instrumentation for highly conductive media.

The CRIM model was evaluated for two further matrices in addition to water wet soil. This was not possible with the Topp model as this relation is only valid for a matrix of water wet soil. Predictions based on the CRIM model had an average error of between 3.77% and 5.25% for the soil-oil and sand-water experiments respectively compared to measured data. As a result, the CRIM model will be implemented in COMSOL multiphysics to couple the electrostatic simulation to the soil transport simulation to allow for multiphysics simulations.

Experiments have shown that the temperature dependency of the dielectric constant of water will not contribute substantial error over the normal operating temperature range of the instrument. However temperature correction may be easily integrated using a simple mathematical model [35].

Ultimately, for further evaluation of the simulation it is felt that a higher frequency range will improve capacitance measurement capabilities for water-wet soils. Typically soil dielectric measurement instruments operate in the 10-200 MHz region [13] however the current instrumentation is limited to 1 MHz. It is anticipated that a measurement range up to 75 MHz will be sufficient [33].

4.2. Future work

Some time must be spent to further understand the limitations of the measurement instrument. Ideally a calibration curve may be generated using complex impedance measurements, allowing the correction of measurement data and hence extending the measurement range of the instrument.

Further work will be carried out to greater parameterise the soil samples. In doing so any uncertainty regarding the current porosity value will be minimised, this will allow a better comparison between the Topp and CRIM models.

The current measurement vessel will be replaced with one allowing the spatial moisture distribution to be mapped, i.e. a vessel with a larger number of independent electrodes.

Fluid flow in a porous media and electrostatics models have been implemented in isolation, this paper focused only on the latter. These models will be coupled through the dielectric constant vs. soil saturation curve as predicted by a mixture model. This will allow the simulation of electric fields due to varying soil water distributions hence facilitating the implementation of a multiphysics based forward solver in COMSOL multiphysics. The forward solver will be optimised for typical tomographic vessel geometries and will allow the generation of calibration curves in order to correct for measurement error. Ultimately a fully integrated electrostatics model based on the fluid flow as predicted by the Richards equation will be developed.

In the long term, the soil retention curves will be obtained experimentally and the resulting parameterisation substituted into the Richards equation solver in COMSOL Multiphysics. Root growth models and root water uptake models will be integrated into the multiphysics models. The final, fully integrated model will be used in the reconstruction of ECT images and enable prior information in the inverse problem.

References

[1] Bontemps C and Couture S 2002 Irrigation water demand for the decision maker *Environment and Development Economics* **7** 643–57

[2] Lange M D 2001 Water law and human rights – roles and responsibilities *Water Science and Technology* **43 4** 143–50

[3] Hunt R, Causton D R, Shipley B and Askew A P 2002 A modern tool for classical plant growth analysis *Annals of Botany* **90** 485-88

[4] Leitner D and Schnepf 2009 Root growth simulation using L-systems *Proceedings of Algoritmy* 313–20

[5] Schnepf A and Leitner D 2009 FEM simulation of below ground processes on a 3-dimensional root system geometry using Distmesh and COMSOL Multiphysics, *Proceedings of Algoritmy* 321-30

[6] Lynch J P, Nielsen K L, Davis R D and Jablokow A G 1997 SimRoot: Modelling and visualization of root systems *Plant and Soil* **188** 139–51

[7] Javaux M, Schröder T, Vanderborght J and Vereecken H 2008 Use of a three-dimensional detailed modelling approach for predicting root water uptake *Vadose Zone J.* **7** 1079–88

[8] Šimůnek J, van Genuchten M T and Šejna M 2008 Development and applications of the HYDRUS and STANMOD software packages and related codes *Vadose Zone J.* **7** 587–600

[9] COMSOL: Earth Science Module [cited 2010 Feb 23] Available from: http://www.comsol.com/products/es/

[10] Richards L A 1931 Capillary conduction of liquids through porous mediums *Physics* **1** 318-33

[11] Claude Doussan, Loïc Pagès and Gilles Vercambre 1998 Modelling of the Hydraulic Architecture of Root System: An Integrated Approach to Water Absorption – Model Description *Annals of Botany* **81** 213-23

[12] Zhang L, Tian P, Jin X and Tong W 2010 Numerical simulation of forward problem for electrical capacitance tomography using element-free Galerkin method *Engineering Analysis with Boundary Elements* **34** 477–82

[13] Eller H and Denoth A 1996 A capacitive soil moisture sensor *Journal of Hydrology* **185** 137-46

[14] Seyfried M S, Grant LE, Du E and Humes K 2005 Dielectric Loss and Calibration of the Hydra Probe Soil Water Sensor *Vadose Zone* **4** 1070-79

[15] Bogena H R, Huisman J A, Oberdörster and Vereecken H 2007 Evaluation of a low-cost soil water content sensor for wireless network applications *Journal of Hydrology* **344** 32-42

[16] Dobson M C, Ulaby F T, Hallikainen M T and Mohamed A ER 1985 Microwave dielectric behaviour of wet soil – part II: Dielectric mixing models *IEEE Transactions on Geoscience and Remote Sensing* **GE-23 1** 35-46

[17] Topp G C, Davis J L and Annan A P 1980 Electromanetic determination of soil water content: Measurements in coaxial transmission lines *Water Resources Research* **16 3** 574-82

[18] Whalley W R 1993 Consideration on the use of time-domain reflectometry (TDR) for emasuring soil water content *Journal of Soil Science* **44** 1-9

[19] Seyfried M S and Murdock M D 2003 Measurement of soil water content with a 50-MHz soil dielectric sensor *Soil Sci. Soc. Am. J.* **68** 394-403

[20] Francisca F M and Rinald V A 2003 Complex dielectric permittivity of soil-organic mixtures (20 MHz – 1.3 GHz) *Journal of Environmental Engineering* 347-57

[21] Kowalsky M B and Rubin Y 2004 The use of ground-penetrating radar for characterizing sediments under transient flow conditions *Aquifer Characterization, SEPM Special Publication* **80** 107-27

[22] Noborio K 2001 Measurement of soil water content and electrical conductivity by time domain reflectometry: A review *Computers and Electronics in Agriculture* **31** 213-37

[23] Roth K, Schulin R, Flühler H and Attinger W 1990 Calibration of time domain reflectometry for water content measurement using a composite dielectric approach *Water Resources Research* **26 4** 2267-73

[24] Knight R and Endres A 1990 A new concept in modelling the dielectric response of sandstones: Defining a wetted rock and bulk water system *Geophysics* **55 5** 586-94

[25] Knight R and Abad A 1995 Rock/water interaction in dielectric properties: Experiments with hydrophobic sandstones *Geophysics* **60 2** 431-36

[26] Hector L G and Schultz H L 1936 The dielectric constant of air at radiofrequencies *Physics* **7** 133–36

[27] Nightingale N R V, Szwarnowski S, Sheppard J and Grant E H 1981 A coaxial line cell for measuring the permittivity of medium to high loss liquids in the frequency range 2 to 15 GHz *J. Phys. E: Sci. Instrum.* **14** 156-60

[28] Liebe H J, Hufford G A and Manabe T 1991 A model for the complex permittivity of water at frequencies below 1 THz *International Journal of Infrared and Millimeter Waves* **12 7** 659-75

[29] William A J and Horton R 2004 *Soil Physics 6th Edition* John Wiley and Sons, Inc. Hoboken New Jersey 29-30

[30] Sasal M C, Andriulo A E and Taboada M A 2006 Soil porosity characteristics and water movement under zero tillage in silty soils in Argentinian Pampas *Soil & Tillage Research* **87** 9–18

[31] Kusdiana D and Saka S 2004 Two-Step Preparation for catalyst-free biodiesel fuel production hydrolysis and methyl esterification *Applied Biochemistry and Biotechnology* **113–116** 781-91

[32] Zhang N et al 2004 Simulatious measurement of soil water content and salinity using a frequency-response method *Soil Sci. Soc. Am. J.* **68** 1515–25

[33] Kizito CS et al 2008 Frequency, electrical conductivity and temperature analysis of a low-cost capacitance soil moisture sensor *Journal of Hydrology* **352** 367-78

[34] Robinson M and Dean T J 1993 Measurement of near surface soil water content using a capacitance probe *Hydrological Processes* **7** 77-86

[35] Catenaccio A, Daruich Y and Magallanes C 2003 Temperature dependence of the permittivity of water *Chemical Physics Letters* **367** 669–71

Journal of Physics: Conference Series **255** (2010) 012003 doi:10.1088/1742-6596/255/1/012003

Combination of Maximin and Kriging Prediction Methods for Eddy-Current Testing Database Generation

Sándor Bilicz[1,2], **Marc Lambert**[1], **Emmanuel Vazquez**[3] **and Szabolcs Gyimóthy**[2]

[1] Département de Recherche en Électromagnétisme, Laboratoire des Signaux et Systèmes UMR8506 (CNRS-SUPELEC-Univ Paris-Sud), 91192 Gif-sur-Yvette cedex, France

[2] Budapest University of Technology and Economics, Egry J. u. 18, H-1521 Budapest, Hungary

[3] SUPELEC, 91192 Gif-sur-Yvette cedex, France

E-mail: `sandor.bilicz@lss.supelec.fr`

Abstract. Eddy-current testing (ECT) is a widely used nondestructive evaluation technique. The numerical simulation of ECT methods involves high complexity and computational load. However, one needs reliable solutions (within a reasonable CPU time) for these problems to be able to solve the related inverse problem. One possible approach is to build a configuration-specific *database*, consisting of well-chosen samples (corresponding input data – output signal pairs). Once the database has been constructed, the sought information can be retrieved practically in no time. However, the optimal choice of samples raises complex optimization problems. This paper presents a sampling method which aims to achieve databases being optimal in a certain sense. The goal of our approach is to spread out the output samples in the whole conceivable output domain. The method is formalized as a *maximin* problem which is solved step-by-step using the *kriging* prediction.

1. Introduction

The aim of using databases in nondestructive evaluation can be formalized as follows. Given the electromagnetic phenomena involved in the testing method, one is able to construct its mathematical model and compute approximate results by using a numerical simulator. However, in certain applications, the time consumption of such rigorous simulations is unacceptable. Databases offer an alternative way: build a database (for the given, specific problem), containing pre-calculated results, then, one just recourses to this database and retrieves the sought information. The generation of a database might take long time, but the data retrieval (which is aimed in the application) is then practically a real-time task.

In spite of the simplicity of the basic idea, the database generation raises numerous questions. An exhaustive presentation of the problem along with a database generation algorithm is found in [6].

In this paper, a new approach is proposed for the generation of databases with eddy-current testing applications. The procedure is a combination of a kind of maximin method and the kriging prediction.

2. Basic database concepts

Let us imagine an ECT configuration, in which a defect (e.g., one single void, or several thin cracks close to each other) is characterized by a vector $\mathbf{x}$ of p parameters, $\mathbf{x} = [x_1, x_2, \ldots, x_p]$, the latter being the *input* of the model. The set of all conceivable inputs is referred to as the input domain, being a bounded subset of $\mathbb{R}^p$: $\mathcal{X} \subset \mathbb{R}^p$. Let us denote the *output* vector by $\mathbf{y}$ (e.g., coil impedances at different locations, the length of $\mathbf{y}$ can be several hundred). The set $\mathcal{Y}$ consists of all conceivable $\mathbf{y}$ outputs (corresponding to the inputs $\mathbf{x} \in \mathcal{X}$). Let $||\cdot||_x$ and $||\cdot||_y$ be appropriate norms in $\mathcal{X}$ and $\mathcal{Y}$, thus, a distance is defined on both.

The connection between the input $\mathbf{x}$ and output $\mathbf{y}$ is in principle the physical phenomenon behind the model. Herein this is treated as a "black-box", characterized by the *forward operator* $\mathcal{F}$:

$$\mathbf{y} = \mathcal{F}\{\mathbf{x}\}. \tag{1}$$

$\mathcal{F}$ is deterministic, i.e., knowing the exact input $\mathbf{x}$, the output $\mathbf{y}$ can be expressed exactly as well. The realization of $\mathcal{F}$ is usually based on numerical field computation.

Basically, the database approaches aim to construct a discrete representation of the operator $\mathcal{F}$ by sampling, and by fitting an interpolation method to the samples. The yielded approximation of $\mathcal{F}\{\mathbf{x}\}$ is denoted by $\widetilde{\mathcal{F}}\{\mathbf{x}\}$. A *sample* in the database is a pair of corresponding input-output data: $\{\mathbf{x}_i, \mathbf{y}_i\}$, where $i \in N$ ($N = \{1, 2, \ldots, n\}$ the index set of the database consisting of n samples). The nearest neighbour (NN) method is used for interpolation:

$$\widetilde{\mathcal{F}}\{\mathbf{x}\} = \mathcal{F}\{\mathbf{x}_k\} = \mathbf{y}_k : \ ||\mathbf{x} - \mathbf{x}_k||_x \leq ||\mathbf{x} - \mathbf{x}_j||_x \ \forall j \in N. \tag{2}$$

The precision provided by the interpolation is measured via the error $\varepsilon(\mathbf{x})$:

$$\varepsilon(\mathbf{x}) = \left|\left|\widetilde{\mathcal{F}}\{\mathbf{x}\} - \mathcal{F}\{\mathbf{x}\}\right|\right|_y. \tag{3}$$

The main goal of the database generation strategies is to reduce $\varepsilon(\mathbf{x})$ in a certain sense (e.g., its maximum over $\mathcal{X}$). Obviously, the precision and the number of samples are interacting. Sophisticated sampling methods have already been developed ([5], [6]) but the question of "optimal sampling" is still in the focus of research.

3. The aimed database configuration: output sample spreading

Our aim is to construct databases having approximately uniform sample distribution in the *output* domain $\mathcal{Y}$. In other words, samples are required to fill $\mathcal{Y}$ evenly. This idea of *output sample spreading* has already occurred (in a different application and realization) in [1]. Note that in the domain of "classical" experimental design, several methods are known to fill the *input* domain, but the output domain filling is a different approach.

The expected benefit of such uniform output sampling is easy to understand by considering the forward interpolator (2) and the expression of the error (3). The uniformity of the output sampling provides an output sample "not too far" from each conceivable $\mathbf{y}$. Thus, one can say by intuition that sampling $\mathcal{Y}$ evenly reduces the error of NN interpolation –however, a rigorous proof is not given herein.

4. The proposed generation algorithms: MAXIMIN method

The sought uniform output sample distribution is attempted to achieve by using the *maximin method.* In our context, *maximin* means maximizing the minimal distance between any pair of output samples. In other words, a sample distribution is aimed where the distance between each output sample and its nearest neighbour is as large as possible.

Two slightly different incremental algorithms are proposed to generate databases of this maximin property. For a formal description, let us introduce the so-called *distance functions*:

$$Q_i(\mathbf{x}) = ||\mathcal{F}(\mathbf{x}) - \mathbf{y}_i||_y, \text{ where } i \in N. \tag{4}$$

One distance function is related to each sample, $Q_i(\mathbf{x})$ corresponds to the i-th one. These distance functions play the main role in both algorithms presented in the next subsections.

4.1. Simple approach: only adding samples

The flowchart of the algorithm is shown in Fig. 1. The different tasks are summarized in the following.

- **Initialization:** At the beginning, some initial input samples are chosen. In our experiments, initial input samples are placed at the nodes of a regular grid over $\mathcal{X}$ –as no information about the studied problem is available *a priori.*
- **Simulation:** The forward operator is evaluated to obtain the output samples related to the initial input samples and to the currently added new input samples during the generation process. The forward operator is represented by a mathematical model (based on Maxwell's equations in eddy-current testing) and solved via numerical simulation (e.g. finite elements, integral equations, etc.).
- **New sample:** To add the next, $(n+1)$-th sample to the database, the evaluation of the following expression is needed:

$$\mathbf{x}_{n+1} = \arg\max_{\mathbf{x}\in\mathcal{X}} \left[\min_{i=1,2,\ldots,n} Q_i(\mathbf{x}) \right]. \tag{5}$$

  (5) is an optimization problem, being numerically demanding to solve ($Q_i(\mathbf{x})$ implies the computation of one direct problem). To overcome this difficulty, the distance functions $Q_i(\mathbf{x})$ are replaced by their kriging predictions $\widehat{Q}_i(\mathbf{x})$. For details about kriging, see subsection 4.3.

  To avoid confusion, we have to emphasize that the formula (5) is not the "classical" maximin rule (known in decision theory, game theory, statistics, etc.), as in the classical sense, the "max" and "min" operators operate on the same domain. In our case, "min" is defined pointwise over a discrete set of distance functions, whereas "arg max" seeks for the global maximizer over the continuous input domain.
- **Stop?:** The stopping criterion decides if the database passes a given criterion (for instance, might be related to the interpolation precision in a certain sense or to the size of the database). If the stopping criterion is met, the sample insertion stops and the database is ready.

The sequential sample insertion by using this criterion yields databases being *approximately* optimal in the sense of uniform output sampling. Since the algorithm looks only one-step ahead and no back step is allowed, the output sample distribution cannot be expected to be perfectly uniform. Moreover, the initial samples are not chosen according to the criterion (5).

4.2. Advanced approach: adding and removing samples

The drawbacks of the first approach are intended to be eliminated by allowing the algorithm to remove samples during the generation process. This advanced approach (shown in Fig. 2) consists in two loops: one for building and an other one for decoupling the database in order to improve the uniformity of the output sampling. The tasks of the already presented parts of the flow chart remained the same, only the new parts are discussed below.

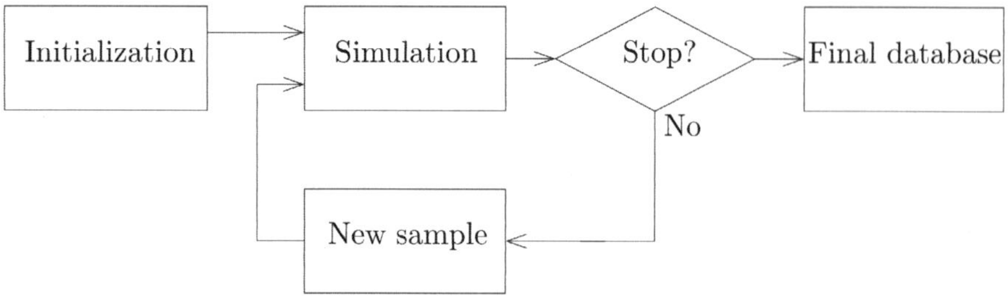

Figure 1: Flow chart of the database generation process without the sample-removing part (*simple approach*).

- **Remove?:** A sample is removed if a given condition is satisfied at this point. Our decision rule is based on the number of cycles: removing is done in every l-th cycle (e.g., $l = 3$ in the example presented hereafter).
- **Remove sample:** The criterion for choosing the sample to be removed consists in two steps. First, the procedure finds the pair of closest output samples $\mathbf{y}_a$ and $\mathbf{y}_b$:

$$||\mathbf{y}_a - \mathbf{y}_b||_y \leq ||\mathbf{y}_i - \mathbf{y}_j||_y \quad \forall i, j \in N, \text{ so that } i \neq j. \tag{6}$$

In the second step, the algorithm chooses between the a-th and b-th samples. If

$$\min_{i \in N \setminus \{a,b\}} ||\mathbf{y}_a - \mathbf{y}_i||_y < \min_{j \in N \setminus \{a,b\}} ||\mathbf{y}_b - \mathbf{y}_j||_y \tag{7}$$

holds, the a-th sample is removed, otherwise the b-th, respectively. In words, the choice depends on the *second* nearest neighbour in $\mathcal{Y}$.

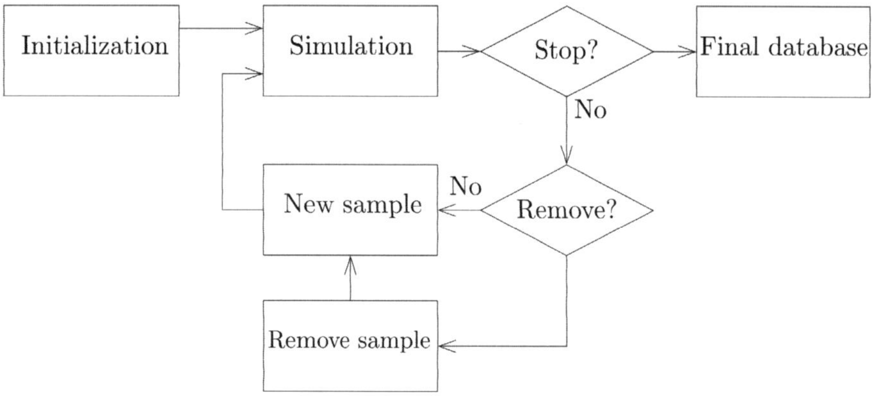

Figure 2: Flow chart of the database generation process with the sample-removing part (*advanced approach*).

4.3. Realization of the method by using kriging

Since neither the forward operator $\mathcal{F}$ nor the distance functions (4) can be expressed analytically, a numerical evaluation (needing the solution of a forward problem) is needed. However, to be able to evaluate the criterion (5), one should know all distance functions over the whole domain $\mathcal{X}$. To overcome this difficulty, *kriging* is used to predict the distance functions at unobserved locations. Kriging is a stochastic tool, developed in the 1960s. Since then several textbooks (for instance, [4]) have dealt with the topic. An ECT-oriented recent presentation is [2]. For the sake of completeness, a brief introduction is given herein as well.

Let us assume that a distance function $Q_i(\mathbf{x})$ is evaluated at the points $\mathbf{x}_1, \mathbf{x}_2, \ldots, \mathbf{x}_n$. The aim of kriging is to find a predictor $\widehat{Q}_i(\mathbf{x})$ of $Q_i(\mathbf{x})$, based on the observations. Kriging models the function $Q_i(\mathbf{x})$ by a Gaussian random process $\xi(\mathbf{x})$, usually assumed to be stationary, i.e., its statistical properties are invariant to shifting in $\mathbf{x}$. Formally, the covariance function $k(\mathbf{h})$ does not depend on $\mathbf{x}$ and the mean is an unknown constant m:

$$k(\mathbf{h}) = \mathrm{cov}\left[\xi(\mathbf{x}), \xi(\mathbf{x}+\mathbf{h})\right] \quad \text{and} \quad \mathbb{E}\left[\xi(\mathbf{x})\right] = m. \tag{8}$$

Kriging computes the Best Linear Unbiased Predictor (BLUP) of $\xi(\mathbf{x})$ in the space spanned by the random variables $\xi(\mathbf{x}_1), \xi(\mathbf{x}_2), \ldots, \xi(\mathbf{x}_n)$. The predictor has the linear form

$$\widehat{\xi}(\mathbf{x}) = \sum_{j=1}^{n} \lambda_j(\mathbf{x})\xi(\mathbf{x}_j). \tag{9}$$

The error of the prediction is $\chi(\mathbf{x}) = \hat{\xi}(\mathbf{x}) - \xi(\mathbf{x})$. The coefficients $\lambda_j(\mathbf{x})$ are chosen so that the mean of $\chi(\mathbf{x})$ is zero (*unbiased* prediction) and the variance of $\chi(\mathbf{x})$ is the smallest possible (*best* prediction). Both requirements are fulfilled provided that the coefficients $\lambda_j(\mathbf{x})$ are the solutions of the linear system of equations:

$$\left[\begin{array}{ccc|c} k(\mathbf{x}_1-\mathbf{x}_1) & \cdots & k(\mathbf{x}_1-\mathbf{x}_n) & 1 \\ \vdots & \ddots & \vdots & \vdots \\ k(\mathbf{x}_n-\mathbf{x}_1) & \cdots & k(\mathbf{x}_n-\mathbf{x}_n) & 1 \\ \hline 1 & \cdots & 1 & 0 \end{array}\right] \left[\begin{array}{c} \lambda_1(\mathbf{x}) \\ \vdots \\ \lambda_n(\mathbf{x}) \\ \hline \mu \end{array}\right] = \left[\begin{array}{c} k(\mathbf{x}-\mathbf{x}_1) \\ \vdots \\ k(\mathbf{x}-\mathbf{x}_n) \\ \hline 1 \end{array}\right], \tag{10}$$

with the Lagrange multiplier μ to enforce the unbiasedness. The covariance function is not known explicitly, but one has to predict it from the observed data. Usually a parameterized class of function is chosen and its parameters are estimated. We only mention herein (as this is not in the scope of the paper) that our covariance functions are chosen from the *Matérn-class* and their parameters are estimated by using a Maximum Likelihood Method.

Once the coefficients $\lambda_j(\mathbf{x})$ are computed by solving (10), the distance function $Q_i(\mathbf{x})$ can be predicted as well:

$$\widehat{Q}_i(\mathbf{x}) = \sum_{j=1}^{n} \lambda_j(\mathbf{x})Q_i(\mathbf{x}_j). \tag{11}$$

However, not only one distance function but all the n ones have to be predicted. A favorable property of the kriging prediction is that the same covariance model can be used to model all distance functions, since all of them inherit the same behaviour from $\mathcal{F}$. Moreover, the observation points can easily be chosen the same for all distance functions. Let us define the distance matrix $\mathbf{Q}$ as

$$\mathbf{Q} = \begin{bmatrix} Q_1(\mathbf{x}_1) & \cdots & Q_1(\mathbf{x}_n) \\ \vdots & \ddots & \vdots \\ Q_n(\mathbf{x}_1) & \cdots & Q_n(\mathbf{x}_n) \end{bmatrix}. \tag{12}$$

The entries of $\mathbf{Q}$ are easy-to-compute, as $Q_i(\mathbf{x}_j) = ||\mathbf{y}_i - \mathbf{y}_j||_y$. By using $\mathbf{Q}$, the prediction (11) can be extended for all distance functions as

$$\left[\widehat{Q}_1(\mathbf{x}), \widehat{Q}_2(\mathbf{x}), \ldots, \widehat{Q}_n(\mathbf{x})\right]^{\mathsf{T}} = \mathbf{Q}\left[\lambda_1(\mathbf{x}), \lambda_2(\mathbf{x}), \ldots, \lambda_n(\mathbf{x})\right]^{\mathsf{T}}. \tag{13}$$

5. Illustrative examples

The presented methods are illustrated via a simple ECT configuration. A non-magnetic metal plate is affected by cuboid-shaped voids, and the impedance variation ΔZ of an air-cored pancake-type coil is measured over a surface above the damaged zone. Numerical data are found in Fig. 3. The surface scan consists of $q = 25 \times 33 = 825$ coil locations on a regular grid centered on the defect. The two input parameters are the length (L) and the depth (D) of a thin (but still volumetric, with a known width) void. The defect opens to the plate surface farther from the coil (OD type). The input parameters are in the ranges

$$0.5\,\text{mm} \leq L \leq 3.5\,\text{mm} \quad \text{and} \quad 10\,\% \leq D \leq 90\,\%. \tag{14}$$

Plate thickness	1.25 mm
Conductivity	10^6 S/m
Coil type	JSAEM-benchmark
Lift-off	0.5 mm
Frequency	150 kHz
Skin-depth	1.30 mm

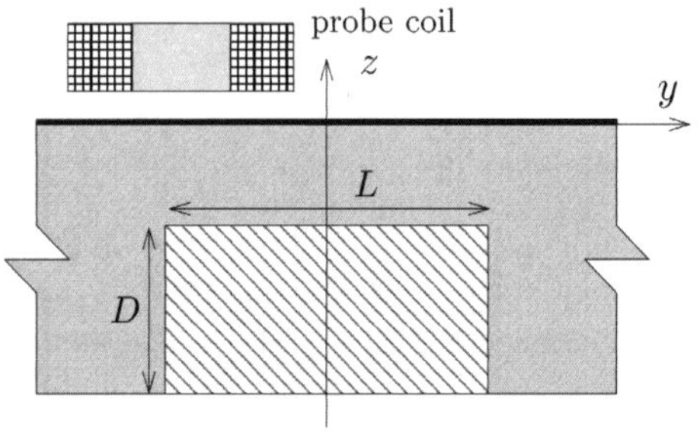

Figure 3: Numerical data and the sketch of the ECT configuration (cross-section).

Synthetic data are used in the examples, the classical volume integral approach [3] is applied to compute the variation of coil impedance due to the defect –the *simulation* box of the flow charts (Fig. 1 and Fig. 2) involves the solution of an integral equation by using the Method-of-Moments.

The presented algorithm can treat high-dimensional output data, thus, it would be possible to define the output of the ECT experiment as a vector containing all $25 \times 33 = 825$ complex impedance values, i.e., $\mathbf{y} = [\Delta Z_1, \Delta Z_2, \dots, \Delta Z_{825}]$. However, the visualization of the output samples would be difficult in this case. This is why –as an illustration– the output signal space is then characterized using only two parameters defined as

$$\text{Mean} = \frac{1}{q}\sum_{j=1}^{q} |\Delta Z|_j \quad \text{and} \quad \text{Shape} = \frac{|\Delta Z|_{center}}{|\Delta Z|_{max}}, \tag{15}$$

where "center" refers to the coil location at the midpoint of the surface scan (also centered on the defect), "max" is related to the largest magnitude of the impedance signal over the whole scan. Thus, the output of the experiment is a 2-element vector: $\mathbf{y} = [\text{Mean}, \text{Shape}]$. Moreover, both output parameters are linearly scaled to the $[0, 1]$ interval a posteriori, for the sake of simplicity.

Both database generation approaches are started with the same initialization part. Initial input samples are spaced at the nodes of a regular, 7×7 grid in the input domain. The initial configuration is presented in Fig. 5. One can see that the initial database does not have uniform output sample distribution: some parts are oversampled whereas others are undersampled in $\mathcal{Y}$.

The final databases generated by the two different methods (and the evolution of the sampling in the electronic version of the paper) are presented in Fig. 6. The stopping criterion was the number of cycles: 100 iterations were allowed in both cases.

One can see in Fig. 6 that the simple approach yields less uniform sample distribution on the output domain. The origin of the distorted distribution is twofold. First, the initial samples (as

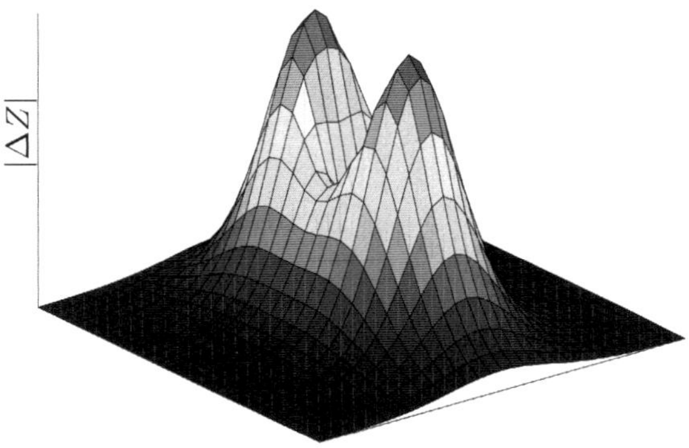

Figure 4: Typical shape of the probe-coil impedance signal on a surface scan centered on a symmetrical defect.

it is shown in Fig. 5) are badly distributed on $\mathcal{Y}$. Second, one also has to keep in mind that not the exact criterion (5) is evaluated, but the distance functions $Q_i(\mathbf{x})$ are approximated by $\widehat{Q}_i(\mathbf{x})$, according to (13). As it is clear in Fig. 6, both disadvantages can be eliminated by the advanced approach, by allowing the sample removal. The advanced approach seems to converge to a quite uniform output sample distribution.

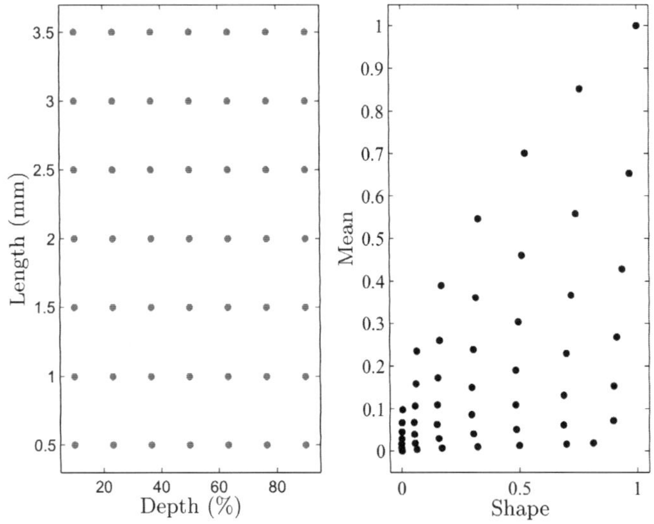

Figure 5: Initial samples. Input domain on the left side, output domain on the right side.

6. Conclusion and future work

A database generation method was presented and illustrated by an eddy-current testing application. The method aims to achieve a uniform output sample distribution in order to provide high interpolation precision by using a relatively small number of samples. Two slightly

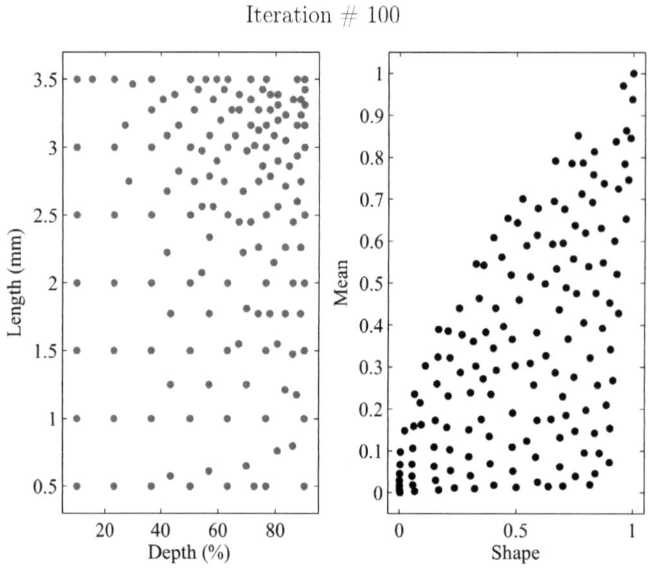

(a) *Simple approach* (no sample removing). See animated movie SamplingOnlyInsertion.mov which presents the evolution of the sampling.

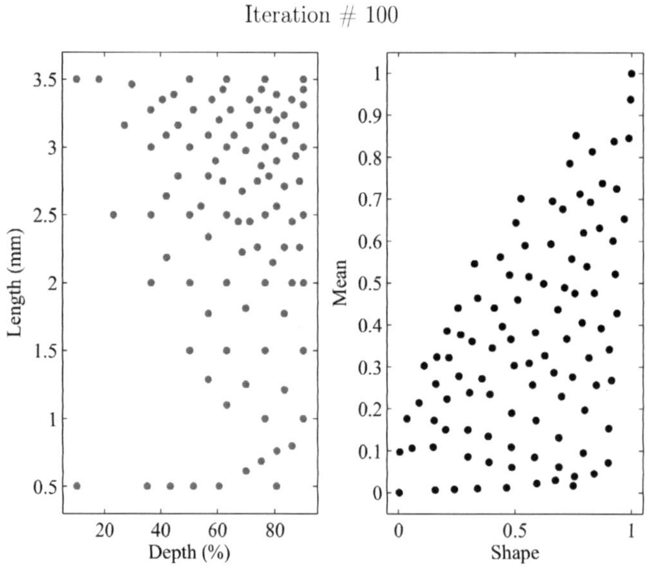

(b) *Advanced approach* (adding and removing samples). See animated movie SamplingWithRemoving.mov which presents the evolution of the sampling.

Figure 6: The databases yielded by the sampling methods after 100 cycles. Input domain on the left side, output domain on the right side.

different implementations were presented: an algorithm which only adds samples one-by-one and an other one which also removes "bad" samples during the generation process. The example illustrated that both algorithms work as they were expected. The advanced sampling approach is believed to converge to a uniform output sample distribution.

However, several questions arise concerning the presented work. The stopping criterion of the generation process should be related to the precision provided by the database. Also the sample removing strategy still needs some further considerations. First, by having a look at Fig. 6b, the *input* sampling appears to be quite coarse in some regions. Obviously, the aimed uniform output sampling might need such distorted input sampling. However, in a practical sense, the kriging prediction of the distance functions (over the input domain $\mathcal{X}$) might be imprecise in the undersampled input regions. Second, the simulation of the forward problem is time consuming. By removing a sample, valuable information is lost. A possible technical improvement is to use "hidden" samples as well, which are not included in the database generated for the end-user but involved only in the kriging model. For instance, all initial samples (over a regular grid in $\mathcal{X}$) can be hidden, and whenever a sample is sentenced to removal in the advanced approach, it becomes hidden instead of deleted.

Beyond the mere interpolation, the *structure* of such an adaptive database provides meta-information about the studied problem as well. Provided that the output sample distribution is uniform, the probability density function (PDF) of the input data can be predicted. Once the PDF is estimated, the solution of the inverse problem in a statistical framework is possible. The authors have not focused on this part yet.

So far only the NN interpolation has been used. A promising idea is to consider more sophisticated interpolators as well. For instance, kriging might also be used for this purpose. However, as the databases should be adapted to the interpolator, the uniform output sampling would not necessarily be the optimal solution in this case. Adapting the database (and developing such sampling strategies) for different interpolators is a challenging open question.

So, the research in the domain of generation and use of databases in eddy-current testing is far from finished. Generalizing the method to other model-based engineering problems has to be considered as well. In the authors' opinion, the results obtained so far are promising and the arising questions justify further studies.

Acknowledgments
The research is partially supported by the Research and Technology Innovation Fund of the Hungarian Government in the frame of the French-Hungarian Bilateral Intergovernmental S&T Cooperation (FR-1/2008) and by DIGITEO cluster's project.

References
[1] R. Bettinger, P. Duchene, L. Pronzato, and E. Thierry. Design of experiments for response diversity. *Journal of Physics: Conference Series*, 135:012017 (8pp), 2008.

[2] S. Bilicz, E. Vazquez, M. Lambert, Sz. Gyimóthy, and J. Pávó. Characterization of a 3D defect using the expected improvement algorithm. *COMPEL*, 28(4):850–863, 2009. DOI 10.1108/03321640910958964.

[3] J.R. Bowler and S. Jenkins. Eddy-current probe impedance due to a volumetric flaw. *Journal of Applied Physics*, 70(3):1107–1114, February 1991.

[4] J.P. Chiles and P. Delfiner. *Geostatistics, Modeling Spatial Uncertainty.* Wiley, 1999.

[5] Sz. Gyimóthy, I. Kiss, and J. Pávó. Adaptive sampling technique based on moving meshes for building data-equidistant inversion databases for NDT. *International Journal of Applied Electromagnetics and Mechanics*, 30.(3-4):309–319, 2009.

[6] J. Pávó and Sz. Gyimóthy. Adaptive inversion database for electromagnetic nondestructive evaluation. *NDT&E International*, 40(3):192–202, 2007.

2.5D inversion of CSEM data in a vertically anisotropic earth

Christophe Ramananjaona and Lucy MacGregor

OHM Rock Solid Images, The Technology Centre, Claymore Drive, Bridge of Don, Aberdeen, AB23 8GD, UK.

E-mail: christophe.ramananjaona@ohmrsi.com

Abstract. The marine Controlled-Source Electromagnetic (CSEM) method is a low frequency (diffusive) electromagnetic subsurface imaging technique aimed at mapping the electric resistivity of the earth by measuring the response to a source dipole emitting an electromagnetic field in a marine environment. Although assuming isotropy for the inversion is the most straightforward approach, in many situations horizontal layering of the earth strata and grain alignment within earth materials creates electric anisotropy. Ignoring this during interpretation may create artifacts in the inversion results. Accounting for this effect therefore requires adequate forward modelling and inversion procedures. We present here an inversion algorithm for vertically anisotropic media based on finite element modelling, the use of Fréchet derivatives, and different types of regularisation. Comparisons between isotropic and anisotropic inversion results are given for the characterisation of an anisotropic earth from data measured in line with the source dipole for both synthetic and real data examples.

1. Introduction

The controlled-source electromagnetic method (CSEM) has proven to be an important addition to traditional seismic imaging techniques in hydrocarbon exploration. Other electromagnetic subsurface imaging methods, such as magnetotellurics and electrical tomography, have been widely used in geophysical exploration. In comparison, CSEM uses a horizontal electric dipole and has the advantage that it is sensitive to both galvanic and inductive effects at the same time. Although assuming isotropy can be sufficient in a number of situations, the presence of electric anisotropy in the earth's crust, due to thin layer interbedding or grain alignments in the sediments can significantly alter the response measured at the seafloor [1]. Isotropic inversions of CSEM data characterising such structures often provide oscillatory artificial results [2] or results which are not consistent with other subsurface imaging techniques (well logging or seismic). As a consequence, accounting for anisotropic effects becomes a necessary challenge to explaining results that isotropic models cannot.

In this paper, we focus on the modelling and inversion of the 3D electromagnetic field measured over a 2D vertically anisotropic earth, also called vertically transverse isotropic earth, and present case study examples of inversions using synthetic as well as real survey data.

2. Modelling

Modelling anisotropy of the earth for CSEM problems has attracted more attention in recent years [3, 4], and methods for interpreting CSEM data using vertically anisotropic models were also recently introduced [5, 2, 6, 7, 8].

We present here a numerical model based on nodal finite elements, which is an extension to include vertical anisotropy of the isotropic model developed by Unsworth *et al.* [9]. The electromagnetic field is decomposed into a primary field $\mathbf{E}^p$ and secondary fields $\mathbf{E}^s$ and $\mathbf{B}^s$, the former being the semi-analytic solution of an isotropic half-space model of conductivity σ_0 and the latter a finite elements solution. The conductivity tensor is defined in an orthonormal coordinate system (O, x, y, z), where O is the origin of the coordinate system, x the invariant (i.e. strike) horizontal direction and z is the vertical direction pointing upwards:

$$\bar{\bar{\sigma}} = \begin{pmatrix} \sigma_h & 0 & 0 \\ 0 & \sigma_h & 0 \\ 0 & 0 & \sigma_v \end{pmatrix} = \begin{pmatrix} \sigma_0 & 0 & 0 \\ 0 & \sigma_0 & 0 \\ 0 & 0 & \sigma_0 \end{pmatrix} + \begin{pmatrix} \Delta\sigma_h & 0 & 0 \\ 0 & \Delta\sigma_h & 0 \\ 0 & 0 & \Delta\sigma_v \end{pmatrix}.$$

The (Ox) direction can be considered invariant if the earth properties can be approximated by infinite bodies in that direction. The secondary electromagnetic field is described by the Maxwell-Faraday and Maxwell-Ampère equations using the quasi-static approximation:

$$\nabla \times \mathbf{E}^s = -\frac{\partial \mathbf{B}^s}{\partial t},$$

$$\nabla \times \mathbf{B}^s = \mu_0 \bar{\bar{\sigma}} \mathbf{E}^s + \mu_0 \overline{\overline{\Delta\sigma}} \mathbf{E}^p.$$

We introduce the anisotropy ratio $\lambda = \sqrt{\frac{\sigma_h}{\sigma_v}}$, and apply a spatio-temporal Fourier transform to Maxwell's equations in the diffusive regime to obtain two PDEs for $\mathbf{E}^s$ and $\mathbf{B}^s$:

$$\begin{aligned} &\mu_0 \nabla . \left[\begin{pmatrix} \frac{\sigma_h}{\gamma_h^2} & 0 \\ 0 & \frac{\sigma_v}{\gamma_v^2} \end{pmatrix} \nabla \hat{E}_x^s \right] - \mu_0 \sigma_h \hat{E}_x^s = \mu_0 \Delta\sigma_h \hat{E}_x^p + \mathrm{i} k_x \mu_0 \nabla . \left[\begin{pmatrix} \frac{\Delta\sigma_h}{\gamma_h^2} & 0 \\ 0 & \frac{\Delta\sigma_v}{\gamma_v^2} \end{pmatrix} \begin{pmatrix} \hat{E}_y^p \\ \hat{E}_z^p \end{pmatrix} \right] \\ &- \mathrm{i} k_x \nabla \times \left[\begin{pmatrix} \frac{1}{\gamma_v^2} & 0 \\ 0 & \frac{1}{\gamma_h^2} \end{pmatrix} \nabla \hat{B}_x^s \right] . \mathbf{u}_x \end{aligned} \quad (1)$$

and

$$\begin{aligned} &\mathrm{i}\omega \nabla . \left[\begin{pmatrix} \frac{1}{\gamma_v^2} & 0 \\ 0 & \frac{1}{\gamma_h^2} \end{pmatrix} \nabla \hat{B}_x^s \right] - \mathrm{i}\omega \hat{B}_x^s = -\mathrm{i}\omega \nabla \times \left[\begin{pmatrix} \frac{\Delta\sigma_h}{\gamma_h^2} & 0 \\ 0 & \frac{\Delta\sigma_v}{\gamma_v^2} \end{pmatrix} \begin{pmatrix} \hat{E}_y^p \\ \hat{E}_z^p \end{pmatrix} \right] . \mathbf{u}_x \\ &- \mathrm{i} k_x \nabla \times \left[\begin{pmatrix} \frac{1}{\gamma_h^2} & 0 \\ 0 & \frac{1}{\gamma_v^2} \end{pmatrix} \nabla \hat{E}_x^s \right] . \mathbf{u}_x \end{aligned} \quad (2)$$

where μ_0 is the magnetic permeability of vacuum, ω the angular frequency, k_x the spatial wavenumber of the Fourier domain, $\gamma_h^2 = k_x^2 - \mathrm{i}\mu_0\omega\sigma_h$, $\gamma_v^2 = k_x^2 - \mathrm{i}\mu_0\omega\sigma_v$ and $\mathrm{i} = \sqrt{-1}$. Equations (1) and (2) are two coupled Sturm-Liouville equations. They admit a unique solution and are simultaneously solved using a finite element discretisation of the two-dimensional domain $D \subset (O, y, z)$ and homogeneous Dirichlet boundary conditions at the edge of the discretisation domain. Then, the different components of the electromagnetic field can be derived from $\hat{E}_x^s$ and $\hat{B}_x^s$ using projections of Maxwell equations onto the coordinate axes [9]. The finite element solution can be computed on an unstructured discretisation mesh, which also enables inversions models including varying seawater depth [10] or complex sub-surface stratigraphy. Due to the high difficulty of estimating the accuracy of the result via functional analysis [11], the accuracy of the code is validated for a 1D model with a flat seafloor against a semi-analytic modelling code using integral equations.

A illustration of 2.5D forward modelling validation is given for the synthetic 1D model in Figure 1. The model consists of a 5000 m deep seawater column of resistivity 0.3125 Ωm

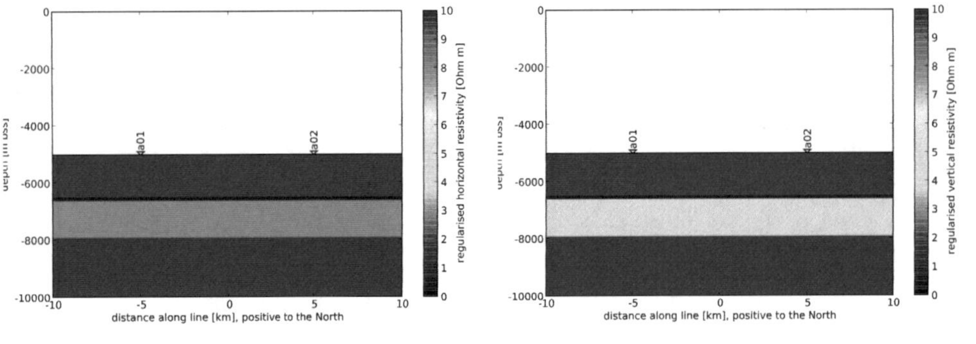

(a) Horizontal resistivity (b) Vertical resistivity

Figure 1. Synthetic 1D vertically anisotropic model with two sources used to validate the finite element code.

and three layers of resistivity 2 Ωm, 50 Ωm and (3, 5) Ωm with thicknesses of 1500 m, 100 m and 1400 m respectively. Two sources are transmitting a signal with a frequency of 0.25 Hz. The response and its difference compared to that calculated using a semi-analytic 1D code are displayed on Figure 2. From the lower panel of Figure 2, we observe that there is less than 1%

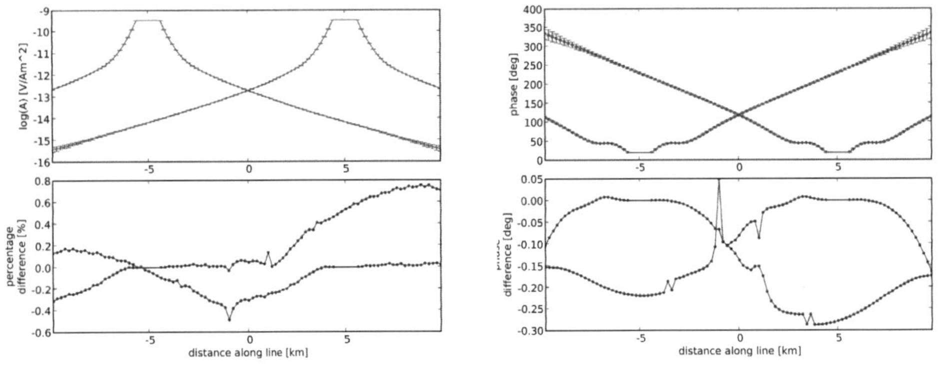

(a) Amplitude (b) Phase

Figure 2. Response and percentage difference between 2.5D finite element modelling and 1D semi-analytic modelling for the model in Figure 1 at 0.25 Hz.

difference at all source-receiver offsets, which indicates that the 2.5D forward modelling code is extremely accurate.

In typical CSEM survey modelling, there are a lot more source positions than receiver positions, since the source dipole continuously emits along the survey tow line, therefore we use the reciprocity theorem of the electromagnetic fields to model sources where the actual receivers are and sample receiver positions along the dipole tow line in order to save computational time.

3. Inversion

The iterative inversion algorithm which is used is an adaptation of the Gauss-Newton method developed by Constable *et al.* [12], also known as the *Occam* inversion. It uses the Fréchet derivatives of a misfit functional characterising the discrepancy between the data and the reconstructed model with respect to the variation of resistivity in a particular cell of the parametrisation domain to reconstruct the two-dimensional resistivity tensor $\bar{\bar{\rho}} = \bar{\bar{\sigma}}^{-1}$. The misfit functional U is given by

$$U = \frac{1}{N} \|Wd - WF(m)\|^2 + \mu \left(\alpha \|\partial m\|^2 + \beta \|m - m^*\|^2\right) - X^*, \tag{3}$$

where d is the data vector, W its associated diagonal weighting matrix (for each datum $d_i \pm \Delta d_i$, $W_i = \frac{1}{\Delta d_i}$), N the number of data, m the vector of model parameters, $F(m)$ the response vector of the model, m^* the *a priori* model, ∂ the smoothing operator, μ its associated Lagrange multiplier, α and β the relative weights of the regularisation terms, and X^* the misfit objective. Here $\|.\|$ denotes the L^2 norm of the measurement space. Typically $d_i = \log_{10} E_{\|,i}$, where $E_{\|,i}$ is the ith measurement of the electric field parallel to the source dipole, so that short and long source-receiver offsets have equal weight in the misfit functional. For the inversions, we use the root mean square (RMS) value of the residuals as indicator of misfit: if $X_i = W_i(d_i - F_i(m))$ is the residual associated to the datum d_i, we have an $RMS = \sqrt{\langle X^2 \rangle}$, where $\langle . \rangle$ denotes the mean. Ideally, the residuals X_i are normally distributed with mean $\langle X \rangle \approx 0$ and standard deviation $\sqrt{\langle X^2 \rangle - \langle X \rangle^2} \approx \sqrt{\langle X^2 \rangle} = RMS$ at the end of the inversion process and therefore the final value of the RMS misfit functional should be close to 1, which should then be the theoretical chosen value for the misfit objective X^*. m is a vector of horizontal and vertical log-resistivities $\log_{10} \rho_h$ and $\log_{10} \rho_v$ and the smoothing term is decomposed as follows:

$$\|\partial m\|^2 = \|\partial_h \rho_h\|^2 + \|\partial_h \rho_v\|^2 + \|\partial_v \rho_h\|^2 + \|\partial_v \rho_v\|^2,$$

where ∂_h and ∂_v respectively represent the horizontal and vertical smoothing matrices.

The misfit functional U with respect to the parameter vector m is locally linearised, so that the search for its minimum provides the update for the parameter vector m at iteration $k+1$ via the equation

$$m_{k+1} = \left[\mu \partial^T \partial + (WJ_k)^T (WJ_k)\right]^{-1} (WJ_k)^T (W\hat{d}_k), \tag{4}$$

with $\hat{d}_k = d - F(m_k) + J_k m_k$ and J_k the Jacobian matrix of derivatives at iteration k defined by the elements $J_{ij,k} = \frac{\partial F_i(m_k)}{\partial m_{j,k}}$ [12]. The Fréchet derivatives of the component electric field parallel to the source dipole in the Fourier domain with respect to the parametric conductivities of a cell l in vertically anisotropic media are given by [13]:

$$\frac{\partial \hat{E}_{\|}}{\partial \sigma_{h,l}} = \int_{D_l} \left(\hat{E}_x \hat{E}_x^* + \hat{E}_y \hat{E}_y^*\right) \mathrm{d}y \mathrm{d}z, \tag{5}$$

and

$$\frac{\partial \hat{E}_{\|}}{\partial \sigma_{v,l}} = \int_{D_l} \hat{E}_z \hat{E}_z^* \mathrm{d}y \mathrm{d}z, \tag{6}$$

where $\hat{\mathbf{E}}^*$ is an adjoint field generated by reciprocal source dipoles, parallel to the real one located at all modelled receivers positions. The adjoint method is proven to be more efficient for computing the derivatives when the number of observation sites is smaller than the number of parameters [14]. The derivatives are then computed at the seafloor in the (x, y, z) domain using an inverse Fourier transform (in practice, a sine or cosine transform). If the source dipole is across strike (i.e. along (Oy)) then the field measured at the seafloor is $E_{\|} = E_y$, whereas if it is along strike (Ox) it is $E_{\|} = E_x$. The Fréchet derivative of the component of the electric field perpendicular to the dipole is similarly computed using an adjoint source dipole perpendicular to the real one.

Although the problem is ill-posed, it can be approximately solved using appropriate regularisation, so that unrealistic results from local minima of the misfit functional are avoided. Possible regularisations are the minimisation of roughness of the model, the minimum norm distance to an *a priori* model [15] or resistivity smoothing breaks at horizons derived from seismic imaging [16].

The electromagnetic field created by a dipole in a conducting medium is characterised by two modes: the Transverse Magnetic (TM) mode, whose current lines are vertical loops, and the Transverse Electric (TE) mode, whose current lines are horizontal loops [13]. Due to the prevalence of the TM mode component of the electric field in the inline configuration, CSEM data acquired in this geometry are primarily sensitive to the vertical resistivity of an anisotropic structure (Figure 3 right panel). The TM mode also provides the primary sensitivity to thin resistive structures in the earth, which are virtually invisible to the inductive TE mode. As a result there is little or no sensitivity to the horizontal resistivity of thin resistivity layers [2], as illustrated in the left panel of Figure 3. As a consequence, inline CSEM measurements will only allow recovery of the vertical resistivity of thin resistive layers, and will provide the best constraint on the vertical resistivity of the overburden. Wide azimuth acquisition including

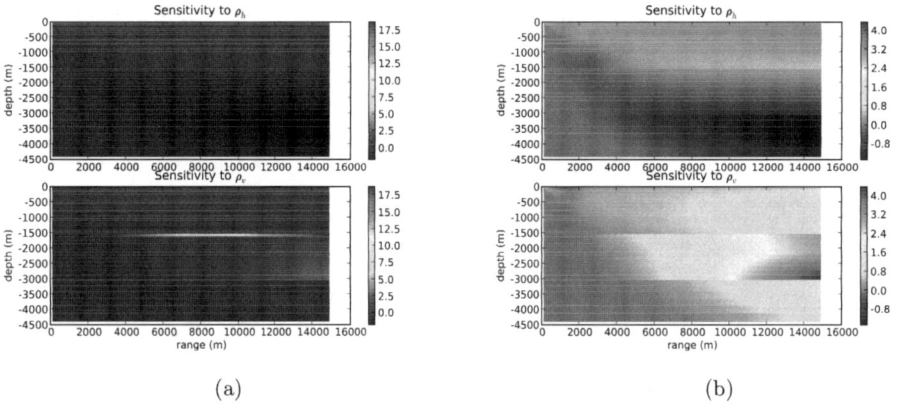

Figure 3. Sensitivity of the inline horizontal electric field to the horizontal (upper panel) and vertical (lower panel) resistivity for the model in Figure 1 (left panel), and the same model without the hydrocarbon reservoir (right panel).

broadside data, where the TE mode is predominant, increases the sensitivity to the horizontal resistivity of thick overburden structures, however the horizontal resistivity of thin resistive layers is still poorly constrained.

4. Examples

4.1. Synthetic data

We present the inversion of synthetic horizontal electric field amplitude and phase data calculated from the model shown in Figure 4. This contains a 2 km wide and 50 m thick hydrocarbon layer of anisotropic resistivity $(50, 60)$ Ωm buried at 900 m below a sloping seafloor. A $(1, 3)$ Ωm anisotropic layer lies above the reservoir and the rest of the model is isotropic. Vertical 1D profiles are given at the abscissa $x = -1000$ m and $x = 1000$ m along the line in Figure 5. The signal is transmitted at 0.125 Hz, 0.375 Hz and 0.625 Hz (corresponding to the fundamental, third and fifth harmonics of a square wave transmission) every 500 m at 40 m above the seafloor, and is measured at ten receivers along a 10 km long line. The initial model for the inversion is an isotropic 2 Ωm half-space for both the isotropic and anisotropic inversions.

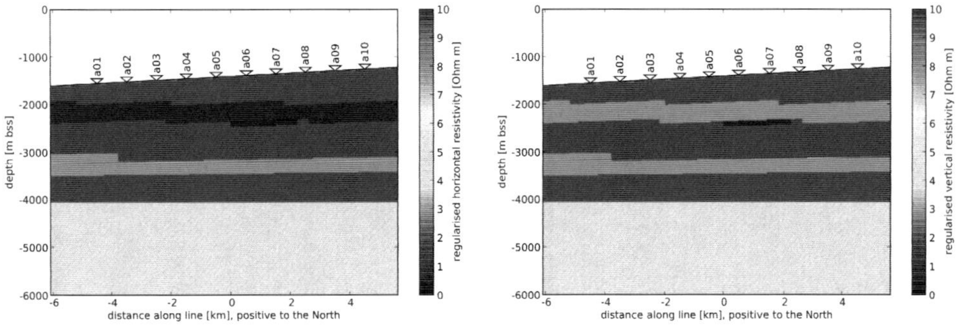

(a) Horizontal resistivity (b) Vertical resistivity

Figure 4. Resistivity structure used in the synthetic inversion study.

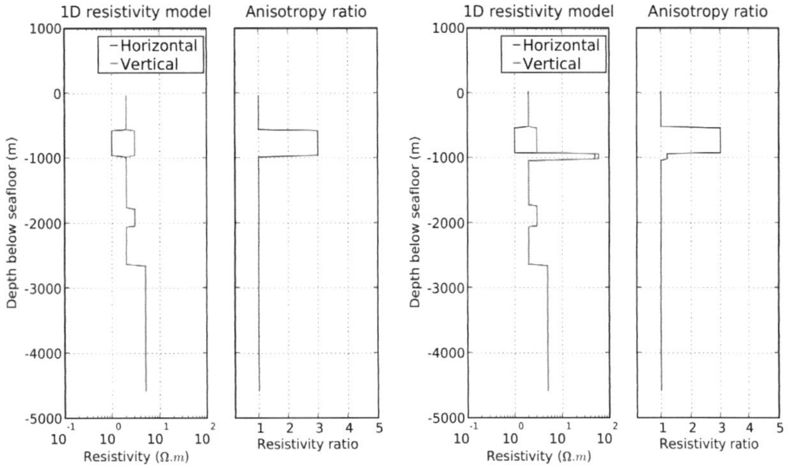

(a) $x = -1000$ m (b) $x = 1000$ m

Figure 5. Vertical 1D profiles with their squared anisotropy ratio λ^2.

The 2D isotropic resistivity section obtained from the isotropic inversion is presented in Figure 6. The residual associated with each data point is plotted as a function of range and distance along the line in Figure 7. The final RMS misfit is 1.081 after 39 iterations. The distribution of residuals is reasonably normally distributed around zero and has no spatial trend: their mean is -0.0159 and their standard deviation is 1.0809. More precisely, the amplitude residuals of the first two frequencies are well distributed around zero, whereas the ones at the third frequency have more negative bias at the shorter ranges. From Figure 6, we observe that the hydrocarbon layer is recovered at the correct lateral location. However its horizontal extent is too large (most likely the result of the isotropic inversion merging the effect of the target and the anisotropic overburden layer) and extra oscillations are included in the overburden layer and below the target. These artifacts are the result of the breakdown of the isotropic assumption.

Conversely, the anisotropic inversion result in Figure 8 does not include such oscillatory resistivity values, and its final RMS misfit is equal to 0.98 after just 9 iterations. The mean of the residuals is -0.0146 and their standard deviation 0.9799. The distribution of residuals is displayed in Figure 9. Vertical 1D profiles through the section are shown in Figure 10. As expected, the horizontal resistivity of the thin hydrocarbon layer is not recovered. However

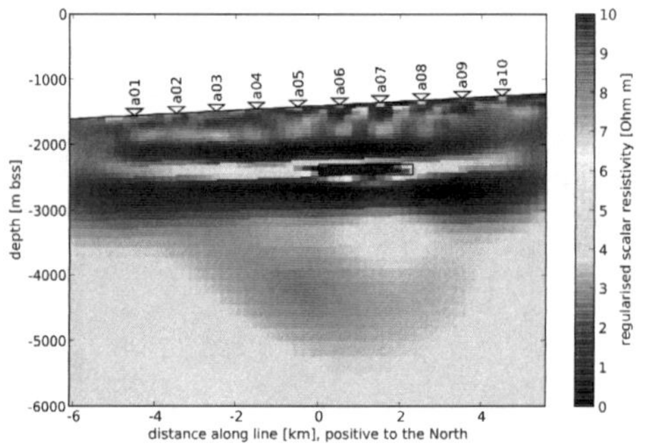

Figure 6. Result of isotropic inversion of synthetic data, iteration 39.

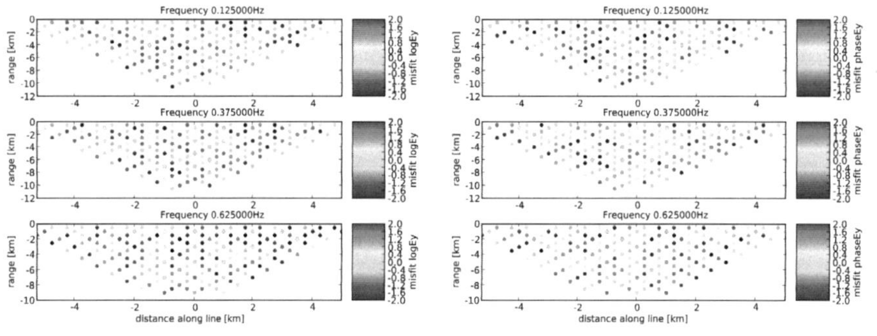

(a) Amplitude (b) Phase

Figure 7. Residuals obtained from the isotropic inversion of synthetic data, RMS=1.081.

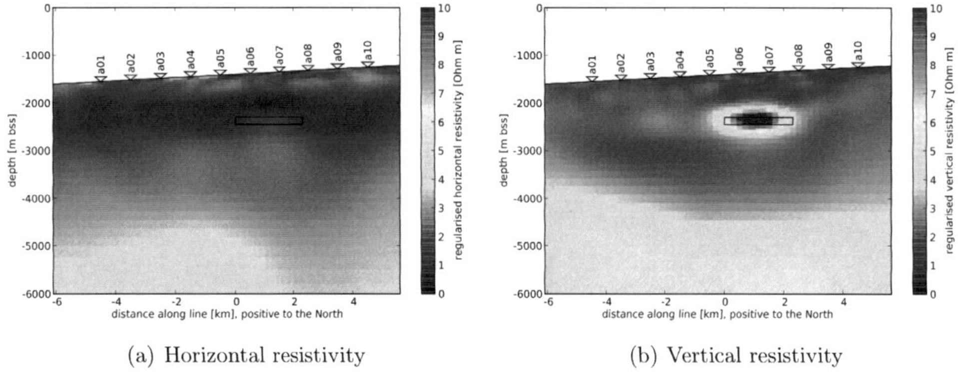

(a) Horizontal resistivity (b) Vertical resistivity

Figure 8. Anisotropic inversion of synthetic data, iteration 9.

the effect of the resistive hydrocarbon layer is clear in the vertical resistivity, although it has been smeared over a range of depths by the smooth regularisation applied. As a result although the resistivity of the hydrocarbon layer as recovered by the inversion is lower than the input value of 60 Ωm, its thickness is greater resulting in the same integrated vertical resistance. The resistivity of the basement of the background model is recovered by both the isotropic and

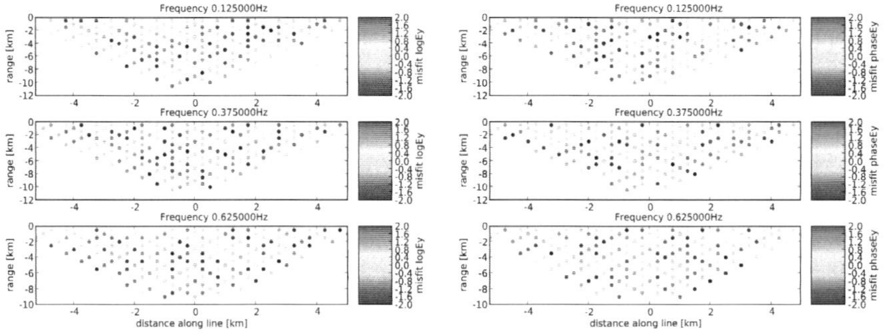

(a) Amplitude (b) Phase

Figure 9. Residuals obtained from the anisotropic inversion, RMS=0.980.

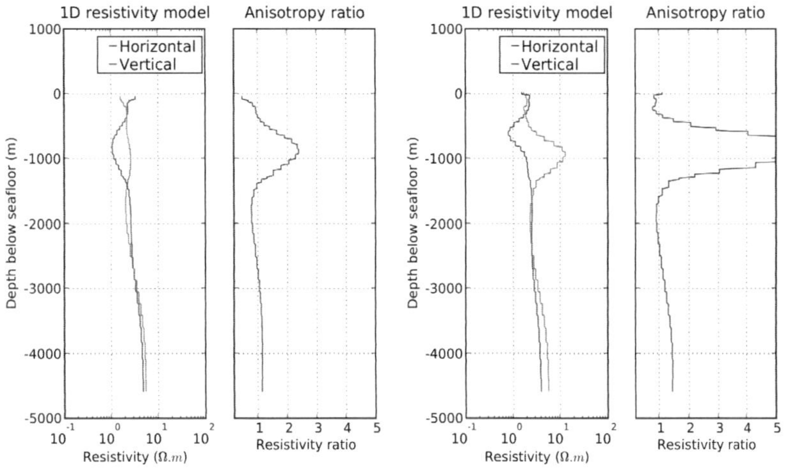

(a) $x = -1000$ m (b) $x = 1000$ m

Figure 10. Vertical 1D profiles with their squared anisotropy ratio λ^2.

anisotropic inversions, but its extent is slightly masked by the presence of the target, especially in the isotropic case. The map of the squared anisotropy ratio λ^2 is displayed in Figure 11. Since

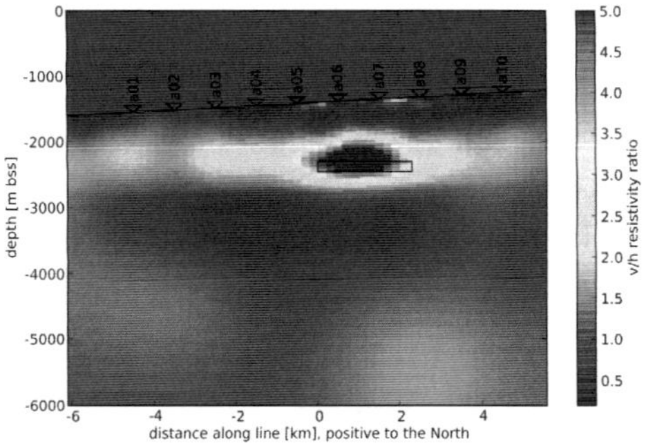

Figure 11. Squared anisotropy ratio for the anisotropic inversion result, iteration 9.

the hydrocarbon reservoir is not recovered in the horizontal resistivity, the ratio is extremely high at this location and obviously does not reflect reality: rather it is caused by the relative sensitivity to the horizontal and vertical resistivity within the layer. However the squared anisotropy ratio of the layer above the target is rather well recovered at ranges where there is sufficient source-receiver coverage: one observes a squared anisotropy ratio above 2, whereas the rest of the reconstructed model is largely isotropic. Away from the resistive reservoir, the maximal squared anisotropy ratio is 2.3 in Figure 10.

4.2. Real data

We now present the result from the inversion of survey data acquired in 2005 offshore West Africa. The survey consisted of two lines crossing a prospective hydrocarbon reservoir lying against a salt diapir (see Figure 12). The structure of the salt diapir is constrained using coincident seismic data. Its top lies at approximately 1.1km below the seafloor, and is expected to be a resistive feature in the earth. In this study we present the results from the East-West trending line, which crosses the salt diapir (labelled line 1). The data consists of the horizontal

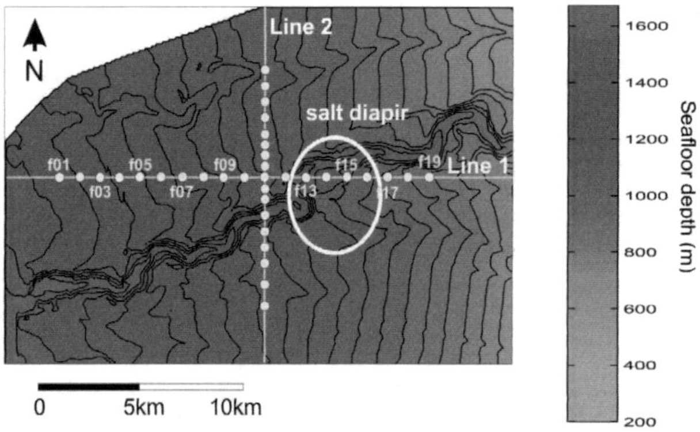

Figure 12. Survey map with the delineation of the salt diapir and the two CSEM lines with the receiver locations.

electric field collected along the 25 km long line at source transmission frequencies of 0.125, 0.375 and 0.625 Hz. An example of the amplitude and phase data collected at receiver f11 (located at the crossing point of the two lines) is shown in Figure 13. Figure 13 also shows the resistivity log from a well located close receiver f11. A standard induction log from a vertical well measures the horizontal component of the resistivity. If the well log resistivity is block averaged, and the CSEM response of this block averaged well log model is compared to the acquired data then the misfit between the two is large. This is because in-line CSEM data, as shown earlier, are primarily sensitive to the vertical resistivity of the earth which is considerably higher. In this case an anisotropic model in which the horizontal resistivity equals that measured in the well, and the vertical anisotropy is one to five times higher provides a much better fit to the data (see Figure 13). The simple 1D modelling of the well log demonstrated that an anisotropic earth is required to explain the data and motivated the need to perform an anisotropic inversion. As a starting point an isotropic inversion of the data was performed, to provide a comparison with the anisotropic results.

The 2D isotropic resistivity section obtained at iteration 10 from the isotropic inversion of the amplitude data from line 1 is presented in Figure 14. The RMS misfit is equal to 0.82, and the spatial distribution of residuals is presented in Figure 15. In this case the misfit is well below 1, because the size of the error bars has been set to a minimum value of 5% of the amplitude, and the inversion succeeded in producing a model which reasonably fits the data to this degree.

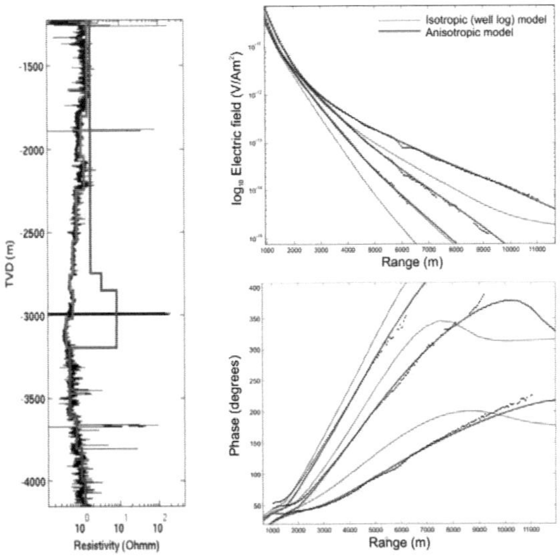

(a) Well models (b) Response

Figure 13. Electric field amplitude and phase data from receiver f11 (left panel). The right panel shows the resistivity log from a well close to this receiver. If a model is constructed from the well log data, which primarily measures the horizontal resistivity, then this does not fit the data (green curve). A much better fit is achieved by an anisotropic model in which the vertical resistivity is considerably higher than the horizontal (red curve)

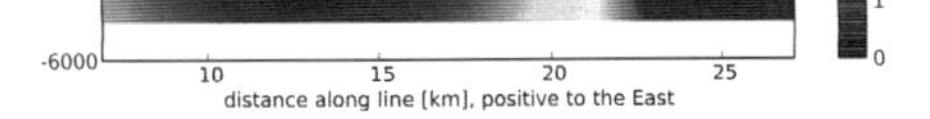

Figure 14. Isotropic inversion result, survey data from line 1, iteration 10.

The absolute values of the error bars have however a minimal impact on the final inversion result when they are modified for the whole dataset. Only a modification of the error bars relating to a part of the dataset can significantly alter the inversion result by altering the data balance between different offsets. The fundamental frequency residuals along the central part of the line are highest, with positive bias at short ranges and negative bias at long range, Their mean is 0.1812. The residuals for the fifth harmonic are also biased, although they are smaller in magnitude with a mean of 0.2553. The resistivity model recovered is shown in Figure 14 and is rather oscillatory. As observed in the synthetic case this roughness is an artifact of the inappropriate isotropic assumption. The increase in resistivity towards of the East of the line may be related to the presence of the salt diapir at depth, however neither the position nor the

Journal of Physics: Conference Series **255** (2010) 012004 doi:10.1088/1742-6596/255/1/012004

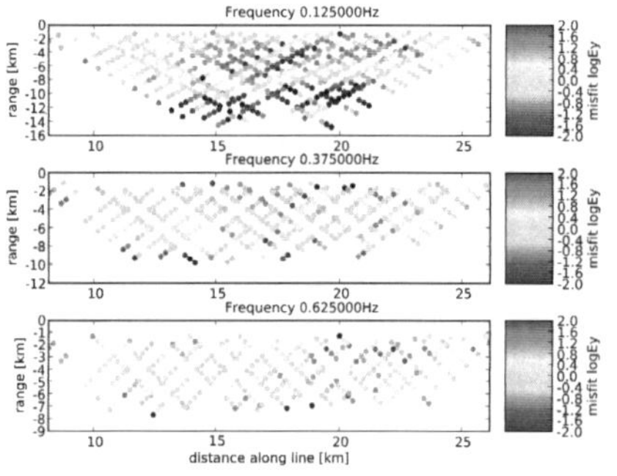

Figure 15. Residuals obtained from the isotropic inversion, RMS=0.82.

resistivity of this are well constrained. In addition, a regional resistive layer is observed at about 3 km below sea surface.

The 2D anisotropic resistivity section obtained at iteration 13 from the anisotropic inversion of the same data is presented in Figure 16. The RMS misfit is equal to 0.38, and the spatial

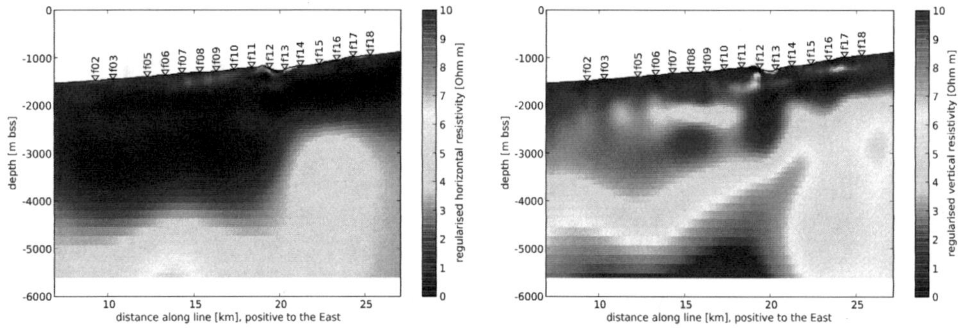

(a) Horizontal resistivity

(b) Vertical resistivity

Figure 16. Anisotropic inversion of survey data from line 1, iteration 13.

distribution of residuals is presented in Figure 17. The final RMS is very low, again due the values assigned to the error bars. The residuals more closely follow a normal distribution this

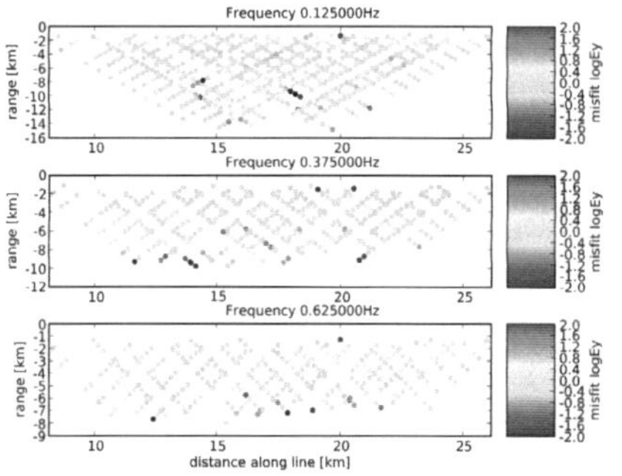

Figure 17. Residuals obtained from the anisotropic inversion, RMS=0.38.

time (their mean is -0.0049 and standard deviation 0.3800), and the RMS misfit is clearly lower than with the isotropic inversion. From Figure 16, we can observe that the salt diapir is clearly resolved in the anisotropic result as an area of elevated horizontal and vertical resistivity. A higher vertical resistivity is also recovered to the West of the diapir in the vicinity of the well. The high horizontal and vertical resistivity below 5000 m indicates electrical basement. The result presented in Figure 16 has a low structural resolution since it results from an unconstrained inversion of the data. One way to improve the resolution would be to include seismically derived structural information, for example breaks in the smoothing constraint at the top basement and diapir interfaces.

The result corendered with the seismic section shows a very good match between the seismic delineation of the salt diapir and the feature in the horizontal resistivity (Figure 18). It can also be noted that the top western part of the diapir is surprisingly more conductive, suggesting the presence of highly saline fluids associated with the diapir in the sub-surface in this area.

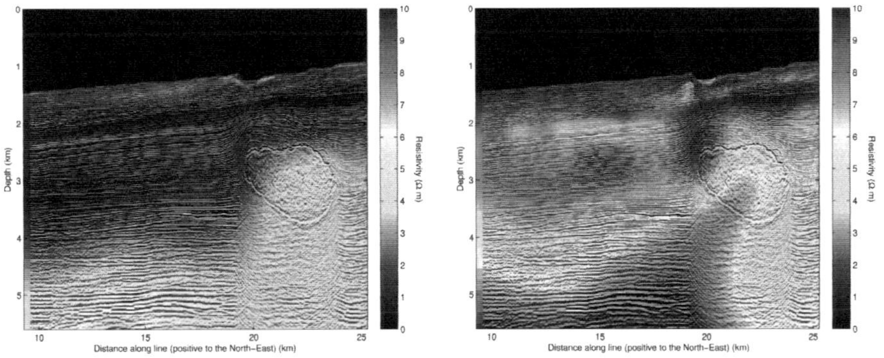

(a) Horizontal resistivity (b) Vertical resistivity

Figure 18. Anisotropic inversion result corendered with the seismic section, iteration 13.

Finally, the 1D profile the anisotropic inversion result is compared with the well resistivity log data measured near receiver f11, i.e. at 18.2 km along the line in Figure 19. The horizontal

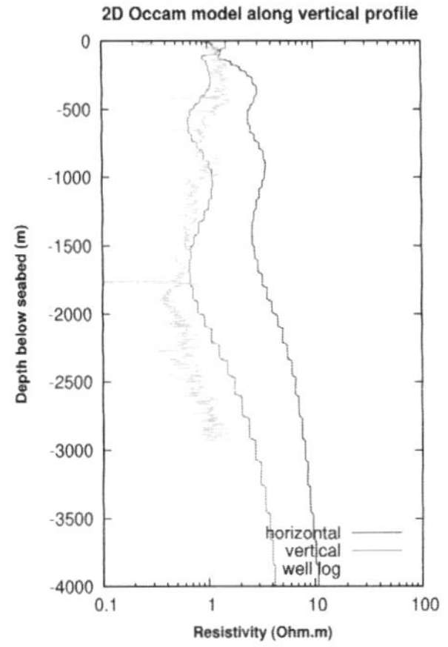

Figure 19. 1D resistivity profile compared to the well log data.

resistivity recovered by the anisotropic inversion agrees with the well log. The reconstructed horizontal resistivity is slightly higher than the well log at depths below 1800 m, also due to the smoothing which spreads the resistivity from highly resistive regions to low resistive ones. The vertical resistivity is higher and follows a trend similar to the horizontal resistivity, with a average squared anisotropy ratio equal to 3.22, consistent with earlier forward modelling.

5. Conclusion

When the earth is anisotropic, an isotropic inversion of inline CSEM data will lead to artifacts in the results which are misleading if interpreted in terms of geology. Comparatively, the anisotropic inversion provides a result which is very close to the true model in synthetic inversions and prevents excessive oscillations in the retrieved resistivity profiles. Another benefit of the anisotropic inversion is that the horizontal resistivity profile recovered from data collected in line with the source dipole will fit a well log (which only measures the horizontal resistivity of the earth), whereas the isotropic inversion will provide a profile which is more likely to look like the vertical resistivity.

Acknowledgements

The authors thank Dana Petroleum for permission to publish the results using the survey data. They also thank Magnus Hagdorn for his useful help during the development of the code.

[1] Tompkins M, Weaver R and MacGregor L 2004 *66th EAGE Conference and Exhibition* (Paris)
[2] Ramananjaona C, Andréis D and MacGregor L 2008 *70th EAGE Conference and Exhibition* (Rome)
[3] Løseth L, Ursin B and Amundsen L 2007 *69th EAGE Conference and Exhibition* (London)
[4] Kong F, Johnstad S, Røsten T and Westerdahl H 2008 *Geophysics* **73** F9–F19
[5] Behrens J P 2005 *The detection of electrical anisotropy in 35 Ma Pacific lithosphere: results from a marine controlled-source electromagnetic survey and implications for hydration of the upper mantle* Ph.D. thesis University of California San Diego
[6] Jing C, Green K and Willen D 2008 *78th SEG Annual Meeting* (Las Vegas)
[7] Carazzone J, Dickens T, Green K, Wahrmund L, Willen D, Commer M and Newman G 2008 *78th SEG Annual Meeting* (Las Vegas)
[8] Shen C and Plessix R 2009 *71st EAGE Conference and Exhibition* (Amsterdam)
[9] Unsworth M J, Travis B J and Chave A D 1993 *Geophysics* **58** 198–214
[10] Hagdorn M, MacGregor L and Robertson B 2008 *70th EAGE Conference and Exhibition* (Rome)
[11] Monk P 2003 *Finite elements methods for Maxwell's equations* (Oxford: Oxford University Press)
[12] Constable S, Parker R and Constable C 1987 *Geophysics* **52** 289–300
[13] Ramananjaona C, MacGregor L and Andréis D 2010 *Geophysical Prospecting* Accepted
[14] MacGillivray P, Oldenburg D, Ellis R and Habashy T 1994 *Geophys. J. Int.* **116** 1–4
[15] Pek J and Santos F A 2006 *Physics of the Earth and Planetary Interiors* **158** 139–158
[16] MacGregor L and Sinha M 2000 *Geophysical Prospecting* **48** 1091–1106

Journal of Physics: Conference Series **255** (2010) 012005 doi:10.1088/1742-6596/255/1/012005

Multiple level-sets for elliptic Cauchy problems in three dimensional domains

A. Leitão, M. Marques Alves

Department of Mathematics, Federal University of St. Catarina, P.O. Box 476, 88040-900 Florianópolis, Brazil

E-mail: acgleitao@gmail.com,maicon@impa.br

Abstract. We analyze a multiple level-set method for solving elliptic Cauchy problems with piecewise constant solutions. This method corresponds to an iterated Tikhonov method for a particular Tikhonov functional based on TV-H^1 penalization. Generalized minimizers for our Tikhonov functional are defined and an existence result is established. Moreover, convergence and stability results of the proposed Tikhonov method are derived. The proposed multiple level-set method is tested numerically, and our experiments demonstrate that the method is able to accurately recover multiple objects as well as multiple contrast levels.

Keywords. elliptic Cauchy problems, level-set methods, multiple levels, Tikhonov regularization

AMS subject classifications. 65J20, 35J60

1. Introduction

The model and the inverse problem Let $\Omega \subset \mathbb{R}^3$, be an open bounded set with piecewise Lipschitz boundary $\partial\Omega$. Moreover, we assume that $\partial\Omega = \overline{\Gamma}_1 \cup \overline{\Gamma}_2$, where Γ_i are two open connected disjoint parts of $\partial\Omega$. We denote by $\mathcal{P}$ the elliptic operator defined in Ω by

$$\mathcal{P}(u) := -\sum_{i,j=1}^{d} D_i(a_{i,j} D_j u)\,, \tag{1}$$

where the real functions $a_{i,j} \in L^\infty(\Omega)$ are such that the matrix $A(x) := (a_{i,j})_{i,j=1}^d$ satisfies $\xi^t A(x)\,\xi > \alpha||\xi||^2$, for all $\xi \in \mathbb{R}^d$ and for a.e. $x \in \Omega$, where $\alpha > 0$.

We denote by *elliptic Cauchy problem* the boundary value problem (BVP)

$$(CP) \qquad \begin{cases} \mathcal{P}u = f\,, \text{ in } \Omega \\ u = g_1\,, \text{ at } \Gamma_1 \\ u_\nu = g_2\,, \text{ at } \Gamma_1 \end{cases},$$

where the functions $(g_1, g_2) \in H^{1/2}(\Gamma_1) \times H_{00}^{1/2}(\Gamma_1)'$ correspond to the problem data, *Cauchy data*, and the real function $f \in L^2(\Omega)$ is a given source term in the model (see [14] for a detailed definition of the Sobolev spaces).

If there exists a distribution $u \in H^1(\Omega)$, which solves the weak formulation of the elliptic equation $\mathcal{P}u = f$ in Ω, and also satisfies the boundary conditions above (at Γ_1) in the sense of the trace operator, we say that u is a (variational) solution of (CP).

Let u be a solution of (CP). It is worth noticing that, if the Neumann trace at Γ_2 of u is known (say $u|_{\Gamma_2} = \varphi$), then u can be computed as a solution of the mixed boundary value problem

$$\mathcal{P}u \;=\; f\,,\ \text{in}\ \Omega \qquad u \;=\; g_1\,,\ \text{at}\ \Gamma_1 \qquad u_\nu \;=\; \varphi\,,\ \text{at}\ \Gamma_2\,,$$

which is a well posed problem in the sense of Hadamard [25]. Therefore, it is enough to consider the task of determining the Neumann trace of u at Γ_2.

Brief overview on elliptic Cauchy problems Elliptic Cauchy problems are not well posed in the sense of Hadamard. A tutorial example given by Hadamard almost 90 years ago during his seminars shows that the solution of (CP) does not depend continuously on the Cauchy data [25, 32]. For a recent analytical investigation of the degree of ill-posedness of elliptic Cauchy problems in two-dimensional bounded Lipschitz domains, we refer the reader to [5].

Existence of solutions for (CP) for arbitrary Cauchy data (g_1, g_2) do not hold. For details we refer the reader to [32, 20]. Actually, one cannot prove existence even in the case of analytical Cauchy data (g_1, g_2) [24].

A given pair of Cauchy data (g_1, g_2) is called *consistent* if the corresponding problem (CP) admits an H^1-solution. It was proved in [3] that the set $M := \{(g_1, g_2) \in H^{1/2}(\Gamma_1) \times [H_{00}^{1/2}(\Gamma_1)]';\ (g_1, g_2)$ consistent Cauchy data$\}$ is a dense subset of $H^{1/2}(\Gamma_1) \times [H_{00}^{1/2}(\Gamma_1)]'$.

What concerns the issue of uniqueness of solutions for (CP), it has been proven that elliptic Cauchy problems admit a unique weak solution in $H^1(\Omega)$ (see, e.g., [20]). A classical uniqueness result for solutions in $C^2(\Omega)$ can be found in [10]. Moreover, a uniqueness result for a class of nonlinear elliptic Cauchy problems can be found in [30].

What concerns numerical investigations of (CP), a large variety of methods can be found in the literature:

M.1) Optimization approach [22, 3];
M.2) Iterative methods [29, 32, 26, 20, 30];
M.3) Backus-Gilbert method [31, 27];
M.4) Optimal control [12];
M.5) Quasi-reversibility [6, 7];
M.6) Level set method [33, 18].

The reason for this strong interest resides on the fact that elliptic Cauchy problems arise in many industrial, engineering and biomedical applications including:

A.1) Expansion of measured surface fields inside a body from partial boundary measurements [3];

A.2) A classical thermostatics problem, which consists in recovering the temperature in a given domain when its distribution and the heat flux are known over the accessible region of the boundary [20];

A.3) The analogous electrostatics case encountered in electric impedance tomography [3];

A.4) Inverse problems related to corrosion detection [2, 33].

Multiple level-set methods for elliptic Cauchy problems Our main goal in this work is to study multiple level-set methods [13, 15] for obtaining regularized solutions of (CP). Multiple level-set approaches for elliptic Cauchy problems are manageable whenever the unknown solution is a simple function defined on Γ_2 assuming at most N different values, i.e. there exists disjoint measurable subsets $D_j \subset \Gamma_2$ and constants c_j, $j = 1 : N$, such that $|\Gamma_2| = \sum_j |D_j|$ and u at Γ_2 is known *a-priori* to satisfy $u_\nu|_{\Gamma_2} = \sum_j c_j \chi_{D_j}$, almost everywhere.

The manuscript is outlined as follows: In Section 2 we write the elliptic Cauchy problem in the functional analytical framework of an (ill-posed) operator equation. This is the starting point for the level set approach derived in the sequel. In Section 3 we investigate a multiple level-set approach for (CP) based on the ideas presented in [15]. First we define a Tikhonov functional related to (CP). This functional is based in the introduction of TV-H^1 penalization. Moreover, we define the concept of generalized minimizers for this functional. Existence of (generalized) minimizers for this Tikhonov functional is proven. Relevant properties of the generalized minimizers as well as properties of the penalization term are investigated. In the sequel we prove convergence and stability results for this Tikhonov regularization method. In Section 4 we introduce a stabilized (smooth) Tikhonov functional. In the main result of this section we prove that the minimizers of the stabilized functional asymptotically approximate the minimizers of the original Tikhonov functional, as the stabilization parameter goes to zero. The corresponding multiple level-set method is derived from an explicit Euler method for solving the evolution equation related to the first order optimality condition of the stabilized functional. Section 5 is devoted to numerics. An efficient implementation of the multiple level-set method is investigated. We are able to improve the performance of the method in [15] by using a specially suited pre-conditioning strategy. Several experiments are provided, in order to illustrate the effectiveness of the multiple level-set method considered in Section 4.

2. Formulation of the inverse problem

We begin by defining the auxiliary problem:

$$\begin{cases} \mathcal{P}v = f \,, \text{ in } \Omega \\ v = g_1 \,, \text{ at } \Gamma_1 \\ v_\nu = \varphi \,, \text{ at } \Gamma_2 \end{cases} . \tag{2}$$

This mixed BVP defines the operator $T : \varphi \mapsto v_\nu|_{\Gamma_1}$. Notice that, if $\varphi = u_\nu|_{\Gamma_2}$, where u is the solution of (CP), then it would follow $T(\varphi) = g_2$. A first least square approach [22] consists in solving the optimization problem

$$\|T(\varphi) - g_2\|^2 \to \min .$$

Due to the superposition principle for linear elliptic BVPs [24], one can split the solution of (2) in $v = v_a + v_b$, where

$$\mathcal{P}v_a = 0 \,, \text{ in } \Omega \qquad v_a = 0 \,, \text{ at } \Gamma_1 \qquad (v_a)_\nu = \varphi \,, \text{ at } \Gamma_2 \,; \tag{3}$$

$$\mathcal{P}v_b = f \,, \text{ in } \Omega \qquad v_b = g_1 \,, \text{ at } \Gamma_1 \qquad (v_b)_\nu = 0 \,, \text{ at } \Gamma_2 \,. \tag{4}$$

Now, we define from (3) the linear operator

$$L : \varphi \mapsto (v_a)_\nu|_{\Gamma_1} \,, \tag{5}$$

and from (4) we define the function $z := (v_b)_\nu|_{\Gamma_1}$. Since $T(\varphi) = L\,\varphi + z$, the Cauchy problem (CP) can be written in the form of the operator equation

$$L\,\varphi \;=\; g_2 - z\,, \tag{6}$$

where the constant term z depends only on the Dirichlet data g_1, on the source term f and on the operator $\mathcal{P}$. Therefore, it can be computed *a-priori*.

In the sequel we shall assume $\Omega \subset \mathbb{R}^3$ and define a functional analytical framework to analyze (6). The Cauchy data is assumed to satisfy

$$(g_1, g_2) \;\in\; H^{1/2}(\Gamma_1) \times H^{1/2}_{00}(\Gamma_1)' \tag{7}$$

and the source term f to be a $L^2(\Omega)$-distribution.

From this choice of g_1 and f, the elliptic theory allow us to conclude that the mixed BVP in (4) has a unique solution $v_b \in H^1(\Omega)$ [14, 24]. Therefore, $z := (v_b)_\nu|_{\Gamma_1} \in H^{1/2}_{00}(\Gamma_1)'$ and the term $g_2 - z$ on the right hand side of (6) is now a distribution in $H^{1/2}_{00}(\Gamma_1)'$.

The next result [18, Proposition 2.1] shows that the linear operator L in (5) is well defined and continuous from $L^{3/2}(\Gamma_2)$ to $H^{1/2}_{00}(\Gamma_1)'$.

Proposition 2.1. *Let $\Omega \subset \mathbb{R}^3$ be defined as in Section 1 and assume the Cauchy data (g_1, g_2) to be given in $H^{1/2}(\Gamma_1) \times H^{1/2}_{00}(\Gamma_1)'$. Then, the operator defined in (5) is an injective bounded linear map $L : L^{3/2}(\Gamma_2) \to H^{1/2}_{00}(\Gamma_1)'$.*

To conclude this section we address an issue related to noisy Cauchy data. If only corrupted noisy data (g_1^δ, g_2^δ) are available for problem (CP), we assume the existence of a consistent Cauchy data (g_1, g_2) satisfying (7) such that

$$\|g_1 - g_1^\delta\|_{L^2(\Gamma_1)} + \|g_2 - g_2^\delta\|_{L^2(\Gamma_1)} \le \delta. \tag{8}$$

Since z in (6) depends continuously on g_1 in the $H^{1/2}(\Gamma_1)$ topology, it is natural to ask whether it is possible to obtain from measured data (g_1^δ, g_2^δ) satisfying (8), a corresponding $z^\delta \in [H^{1/2}_{00}(\Gamma_1)]'$ such that $\|z - z^\delta\|_{[H^{1/2}_{00}(\Gamma_1)]'} \le \delta$. The next result [18, Lemma 2.3] gives a positive answer to this question

Lemma 2.2. *Let the noisy Cauchy data be given as in (8), where $g_1 \in H^s(\Gamma_1)$ for some $s > 1/2$. Then (CP) reduces to the operator equation $L\varphi = g_2^\delta - z^\delta$, where the right hand side satisfies*

$$\|(g_2 - z) - (g_2^\delta - z^\delta)\|_{[H^{1/2}_{00}(\Gamma_1)]'} \le h(\delta). \tag{9}$$

Here $h : \mathbb{R}^+ \to \mathbb{R}^+$ is a function satisfying $\lim_{\delta\to 0} h(\delta) = 0$.

3. A Tikhonov regularization approach

3.1. Generalized minimizers

The starting point of our approach is the assumption that the solution $\overline{\varphi}$ of (6) is a simple function taking only a finite number of possible values. Moreover, we assume the existence of disjoint subsets $D_j \subset \Gamma_2$ and constants c_j, $j = 1 : N$, such that $|\Gamma_2| = \sum_j |D_j|$ and

$$\overline{\varphi} = \sum_{j=1}^{N} c_j \chi_{D_j}, \text{ a.e.} \tag{10}$$

Next we introduce the H^1-functions $\{\phi^j\}_{j=1}^p$, where p is the smallest integer satisfying $2^p \geq N$ (for simplicity of the presentation we shall assume in this paper $N = 4$ and $p = 2$), such that

$$D_1 = \{x \,|\, \phi^1(x) > 0,\, \phi^2(x) > 0\}, \quad D_2 = \{x \,|\, \phi^1(x) > 0,\, \phi^2(x) < 0\},$$
$$D_3 = \{x \,|\, \phi^1(x) < 0,\, \phi^2(x) > 0\}, \quad D_4 = \{x \,|\, \phi^1(x) < 0,\, \phi^2(x) < 0\}.$$

Define $\mathcal{V} = \{u \in L^\infty(\Gamma_2) \,|\, u = \chi_D\,,\, D \subset \Gamma_2 \text{ measurable}, \mathcal{H}^{n-1}(\partial D) < \infty\}$, where $\mathcal{H}^{n-1}$ denotes the $(n-1)$-dimensional Hausdorff-measure, and $q : \mathcal{V} \times \mathcal{V} \to L^\infty(\Gamma_2)$ is given by

$$q(u_1, u_2) = c_1 u_1 u_2 + c_2(1 - u_1)u_2 + c_3 u_1(1 - u_2) + c_4(1 - u_1)(1 - u_2). \tag{11}$$

Set $\mathcal{U} = q(\mathcal{V}, \mathcal{V})$, i.e., $\mathcal{U} = \{u \in L^\infty(\Gamma_2) \,|\, u = q(u_1, u_2),\, u_1, u_2 \in \mathcal{V}\}$, and define the operator $P : H^1(\Gamma_2) \times H^1(\Gamma_2) \ni (\phi^1, \phi^2) \mapsto q(H(\phi^1), H(\phi^2)) \in \mathcal{U}$, where $H : H^1(\Gamma_2) \to \mathcal{V}$, is the Heaviside projector. Using (10) and the above definitions we can represent $\overline{\varphi}$ by

$$\overline{\varphi} = P(\phi^1, \phi^2)\,,$$

and the inverse problem (6) can be written in the form of the operator equation

$$L(P(\phi^1, \phi^2)) = g_2^\delta - z^\delta, \tag{12}$$

with noisy data satisfying (9). Once an approximate solution (ϕ^1, ϕ^2) of (12) is obtained, a corresponding solution of (6) is given by $\overline{\varphi} = P(\phi^1, \phi^2)$.

The multiple level-set method proposed in this paper corresponds to a continuous evolution of the functions ϕ^1 and ϕ^2 for an artificial time t. This evolution aims to minimize the Tikhonov functional

$$\mathcal{G}_\alpha(\phi^1, \phi^2) = \|L\,(P(\phi^1, \phi^2)) - (g_2^\delta - z^\delta)\|_Y^2 + \alpha \sum_{j=1}^{2} \{\beta |H(\phi^j)|_{BV} + \|\phi^j - \phi_0^j\|_{H^1}^2\} \tag{13}$$

based on TV–H^1 penalization. Here $\alpha > 0$ plays the role of a regularization parameter and $\beta > 0$ is a scaling factor. The BV-seminorm terms penalize the length of the Hausdorff measure of the boundary of the set $\{x \,|\, \phi^1(x) \geq 0 \text{ and } \phi^2(x) \geq 0\}$ and play an important role in the analysis of convergence of our regularization procedure.

Since P is discontinuous, one cannot prove that the Tikhonov functional (13) attains a minimizer. In order to guarantee existence of minimizers for $\mathcal{G}_\alpha$, it is necessary to use the concept of generalized minimizers introduced in [15].

Definition 3.1. *Let the boundary part $\Gamma_2 \subset \partial\Omega$ be defined as in Section 1.*

(i) The set Ad of admissible parameters consists of tuples $(z^1, z^2, \phi^1, \phi^2) \in (L^\infty(\Gamma_2))^2 \times (H^1(\Gamma_2))^2$ such that there exist sequences $\{\phi_k^1\}_{k\in\mathbb{N}}$, $\{\phi_k^2\}_{k\in\mathbb{N}}$ in $H^1(\Gamma_2)$ and a sequence $\{\varepsilon_k\}_{k\in\mathbb{N}}$ of positive numbers converging to zero, satisfying[2]

$$\lim_{k\to\infty} \|\phi_k^j - \phi^j\|_{L^2(\Gamma_2)} = 0 \quad \text{and} \quad \lim_{k\to\infty} \|H_{\varepsilon_k}(\phi_k^j) - z^j\|_{L^{3/2}(\Gamma_2)} = 0,\; j = 1, 2.$$

[2] $$H_\varepsilon(\phi^j)(x) := \begin{cases} 0 & \text{for } \phi^j(x) < -\varepsilon, \\ 1 + \frac{\phi^j(x)}{\varepsilon} & \text{for } \phi^j(x) \in [-\varepsilon, 0], \\ 1 & \text{for } \phi^j(x) > 0. \end{cases}$$

(ii) A **generalized minimizer** *of* $\mathcal{G}_\alpha$ *is an admissible parameter* $(z^1, z^2, \phi^1, \phi^2)$ *minimizing*

$$\mathcal{G}_\alpha(z^1, z^2, \phi^1, \phi^2) := \|L\,(q(z^1, z^2)) - (g_2^\delta - z^\delta)\|_Y^2 + \alpha\,\rho(z^1, z^2, \phi^1, \phi^2) \tag{14}$$

over Ad. Here the functional ρ *is defined by*

$$\rho(z^1, z^2, \phi^1, \phi^2) := \inf\Big\{ \liminf_{k\to\infty} \sum_{j=1}^{2} \big(\beta |H_{\varepsilon_k}(\phi_k^j)|_{BV} + \|\phi_k^j - \phi_0^j\|_{H^1}^2\big)\Big\},$$

where the infimum is taken with respect to all sequences $\{\varepsilon_k\}$ *and* $\{(\phi_k^1, \phi_k^2)\}$ *as in 1.*

The set Ad is non empty [15, Remark 2]. Moreover, since the solution of (6) is assumed to satisfy $\overline{\varphi} = P(\phi^1, \phi^2)$, if the functions $\phi^j \in H^1(\Gamma_2)$ are such that $|\nabla\phi^j| \neq 0$ in a neighborhood of the level set $\{\phi^j = 0\}$, we can define the constant sequences $\{\phi_k^j = \phi^j\}_k$ and $z^j = H(\phi^j)$, and estimate

$$\|H_{\varepsilon_k}(\phi_k^j) - z^j\|_{L^{3/2}(\Gamma_2)} \le C \int_{-\varepsilon_k}^{0} 1\, dt \to 0, \quad \varepsilon_k \to 0,$$

where C is a positive number and $\{\varepsilon_k\}$ is any positive sequence converging to zero. Consequently $(z^1, z^2, \phi^1, \phi^2)$ is in Ad and satisfies

$$L(q(z^1, z^2)) = g_2 - z\,. \tag{15}$$

3.2. Convergence analysis

In the next theorem we summarize results on coercivity and lower semicontinuity of the functional ρ and well-posedness of $\mathcal{G}_\alpha$ which can be found in [15].

Theorem 3.1. *Let the functionals* $\rho, \mathcal{G}_\alpha$ *be defined as in item 2 of Definition 3.1. The following assertions hold true:*

(i) The functional $\rho(z^1, z^2, \phi^1, \phi^2)$ *is coercive and strong-weak lowersemicontinuous in the* $BV(\Gamma_2)^2 \times H^1(\Gamma_2)^2$ *and* $L^{3/2}(\Gamma_2)^2 \times H^1(\Gamma_2)^2$ *topology, respectively;*

(ii) The functional $\mathcal{G}_\alpha$ *in* (13) *attains minimizers on the set Ad.*

Proof. See Lemma 4, Lemma 5 and Theorem 6 in [15]. □

In the next theorem, we present the results of convergence and stability of the approximations for the solution of (6). These results were originally proved for a nonlinear inverse problem in [15], where a notion of minimum-norm solution was introduced.

Theorem 3.2. *The following assertions hold true:*

(i) Assume that we have exact data, i.e. $\delta = 0$ *and* $\beta > 0$. *For every* $\alpha > 0$ *let* $(z_\alpha^1, z_\alpha^2, \phi_\alpha^1, \phi_\alpha^2)$ *denote a minimizer of* $\mathcal{G}_\alpha$ *on the set Ad. Then, for every sequence of positive numbers* $\{\alpha_k\}_{k\in\mathbb{N}}$ *converging to zero, there exists a subsequence (denoted again by* $\{\alpha_k\}_{k\in\mathbb{N}}$*) such that* $(z_{\alpha_k}^1, z_{\alpha_k}^2, \phi_{\alpha_k}^1, \phi_{\alpha_k}^2)$ *is strongly convergent in* $(L^{3/2}(\Gamma_2))^2 \times (L^2(\Gamma_2))^2$ *to some limit point* $(z^1, z^2, \phi^1, \phi^2)$. *Moreover, the limit* (z^1, z^2) *is a solution of* (6), *i.e.* $L(q(z^1, z^2)) = g_2 - z$.

(ii) Let the Cauchy data (g_1, g_2) be consistent and $\alpha = \alpha(\delta)$ be a positive function satisfying $\lim_{\delta\to 0}\alpha(\delta) = 0$ and $\lim_{\delta\to 0}\delta^2\alpha(\delta)^{-1} = 0$. Moreover, let $\{\delta_k\}_{k\in\mathbb{N}}$ be a sequence of positive numbers converging to zero and $\{(g_1^{\delta_k}, g_2^{\delta_k})\}_{k\in\mathbb{N}}$ be a corresponding noisy data satisfying (8). Then, there exist a subsequence (denoted again by $\{\delta_k\}$) and a sequence $\{\alpha_k := \alpha(\delta_k)\}_{k\in\mathbb{N}}$ such that $(z^1_{\alpha_k}, z^2_{\alpha_k}, \phi^1_{\alpha_k}, \phi^2_{\alpha_k})$ converges in $(L^{3/2}(\Gamma_2))^2 \times (L^2(\Gamma_2))^2$ to some limit point $(z^1, z^2, \phi^1, \phi^2)$. Moreover, (z^1, z^2) satisfy $L(q(z^1, z^2)) = g_2 - z$.

Proof. The proof follows from the results in [15, Theorems 8 and 9] applied to the operator equation $L(q(z^1, z^2)) = g_2 - z$. □

4. Multiple level-set approximations

4.1. A Stabilized functional

From a numerical viewpoint it is important to define a functional which can be handled numerically. Our aim in this direction is to try to find minimizers which "approximate" the minimizers of $\mathcal{G}_\alpha$ by means of the stabilized functional defined by

$$\mathcal{G}_{\varepsilon,\alpha}(\phi^1, \phi^2) := \|L(P_\varepsilon(\phi^1, \phi^2)) - (g_2^\delta - z^\delta)\|_Y^2 + \alpha \sum_{j=1}^{2} \{\beta |H_\varepsilon(\phi^j)|_{BV} + \|\phi^j - \phi_0^j\|_{H^1}^2\}, \quad (16)$$

where ε, α and β are positive real numbers and $P_\varepsilon(\phi^1, \phi^2) = q(H_\varepsilon(\phi^1), H_\varepsilon(\phi^2))$ is a smooth approximation of the discontinuous operator P.

By Lemma 10 of [15] we have that $\mathcal{G}_{\varepsilon,\alpha}$ is well-posed, i.e., attains minimizers on $(H^1(\Gamma_2))^2$.

The next theorem shows that, for $\varepsilon \to 0$, the minimizers of $\mathcal{G}_{\varepsilon,\alpha}$ approximate a (generalized) minimizer of $\mathcal{G}_\alpha$.

Theorem 4.1. *Let $\alpha, \beta > 0$ be given. For each $\varepsilon > 0$ denote by $(\phi^1_{\varepsilon,\alpha}, \phi^2_{\varepsilon,\alpha})$ a minimizer of $\mathcal{G}_{\varepsilon,\alpha}$. There exists a sequence of positive numbers $\{\varepsilon_k\}$ converging to zero such that $(H_{\varepsilon_k}(\phi^1_{\varepsilon_k,\alpha}), H_{\varepsilon_k}(\phi^2_{\varepsilon_k,\alpha}), \phi^1_{\varepsilon_k,\alpha}, \phi^2_{\varepsilon_k,\alpha})$ converges strongly in $(L^{3/2}(\Gamma_2))^2 \times (L^2(\Gamma_2))^2$ and the limit is a generalized minimizer of $\mathcal{G}_\alpha$.*

Proof. Let $(z^1_\alpha, z^2_\alpha, \phi^1_\alpha, \phi^2_\alpha)$ be a minimizer of $\mathcal{G}_\alpha$ on the set Ad. From Definition (3.1), there exists a sequence $\{\varepsilon_k\}_{k\in\mathbb{N}}$ of positive numbers converging to zero and a corresponding sequence $\{(\phi^1_k, \phi^2_k)\}_{k\in\mathbb{N}}$ in $H^1(\Gamma_2)^2$ satisfying

$$\phi^j_k \to \phi^j_\alpha \quad \text{and} \quad H_{\varepsilon_k}(\phi^j_k) \to z^j_\alpha \quad \text{in} \quad L^{3/2}(\Gamma_2),\ j = 1, 2.$$

From Lemma 3 of [15] we can assume that

$$\rho(z^1_\alpha, z^2_\alpha, \phi^1_\alpha, \phi^2_\alpha) = \lim_{k\to\infty} \sum_{j=1}^{2} \left(\beta |H_{\varepsilon_k}(\phi^j_k)|_{BV} + \|\phi^j_k - \phi^j_0\|^2_{H^1(\Gamma_2)}\right).$$

Let $(\phi^1_{\varepsilon_k}, \phi^2_{\varepsilon_k})$ be a minimizer of $\mathcal{G}_{\varepsilon_k,\alpha}$. The sequences $\{\phi^j_{\varepsilon_k}\}$, $j = 1, 2$ are uniformly bounded in $H^1(\Gamma_2)$. Thus, there are weakly convergent subsequences (denoted again by the same indices) and the weak limits are denoted by $\tilde{\phi}^j$, $j = 1, 2$. Similarly, the sequences $\{H_{\varepsilon_k}(\phi^j_{\varepsilon_k})\}_{k\in\mathbb{N}}$, $j = 1, 2$ are uniformly bounded in $BV(\Gamma_2)$. By the compact Sobolev embedding theorem [1, 21] there exist convergent subsequences (denoted with the same indices) and limits are denoted by $\tilde{z}^j$, $j = 1, 2$. Summarizing, we have

$$\phi^j_{\varepsilon_k} \to \tilde{\phi}^j \quad \text{in} \quad L^2(\Gamma_2) \quad \text{and} \quad H_{\varepsilon_k}(\phi^j_{\varepsilon_k}) \to \tilde{z}^j \quad \text{in} \quad L^{3/2}(\Gamma_2), \quad j = 1, 2,$$

as $k \to \infty$. Thus $(\tilde{z}^1, \tilde{z}^2, \tilde{\phi}^1, \tilde{\phi}^2) \in L^{3/2}(\Gamma_2)^2 \times H^1(\Gamma_2)^2$ is admissible.

Arguing with definition of ρ, as in Lemma 1 of [15] and with the continuity of L, we conclude that

$$\|L(q(\tilde{z}^1, \tilde{z}^2)) - (g_2^\delta - z^\delta)\|_Y^2 = \lim_{k\to\infty} \|L(P_{\varepsilon_k}(\phi^1_{\varepsilon_k}, \phi^2_{\varepsilon_k})) - (g_2^\delta - z^\delta)\|_Y^2,$$

$$\rho(\tilde{z}^1, \tilde{z}^2, \tilde{\phi}^1, \tilde{\phi}^2) \le \liminf_{k\to\infty} \sum_{j=1}^{2} \left(\beta |H_{\varepsilon_k}(\phi^j_k)|_{BV} + \|\phi^j_k - \phi^j_0\|^2_{H^1(\Gamma_2)}\right).$$

Therefore,

$$\begin{aligned}
\mathcal{G}_\alpha(\tilde{z}^1, \tilde{z}^2, \tilde{\phi}^1, \tilde{\phi}^2) &= \|L(q(\tilde{z}^1, \tilde{z}^2)) - (g_2^\delta - z^\delta)\|_Y^2 + \alpha\rho(\tilde{z}^1, \tilde{z}^2, \tilde{\phi}^1, \tilde{\phi}^2) \\
&\le \liminf_{k\to\infty} \mathcal{G}_{\varepsilon_k,\alpha}(\phi^1_{\varepsilon_k}, \phi^2_{\varepsilon_k}) \\
&\le \liminf_{k\to\infty} \mathcal{G}_{\varepsilon_k,\alpha}(\phi^1_k, \phi^2_k) \le \limsup_{k\to\infty} \mathcal{G}_{\varepsilon_k,\alpha}(\phi^1_k, \phi^2_k) \\
&\le \limsup_{k\to\infty} \|L(P_{\varepsilon_k}(\phi^1_k, \phi^2_k)) - (g_2^\delta - z^\delta)\|_Y^2 \\
&\quad + \alpha \limsup_{k\to\infty} \sum_{j=1}^{2} \left(\beta |H_{\varepsilon_k}(\phi^j_k)|_{BV} + \|\phi^j_k - \phi^j_0\|^2_{H^1(\Gamma_2)}\right) \\
&= \|L(q(z^1_\alpha, z^2_\alpha)) - (g_2^\delta - z^\delta)\|_Y^2 + \alpha\rho(z^1_\alpha, z^2_\alpha, \phi^1_\alpha, \phi^2_\alpha) \\
&= \mathcal{G}_\alpha(z^1_\alpha, z^2_\alpha, \phi^1_\alpha, \phi^2_\alpha) = \inf \mathcal{G}_\alpha,
\end{aligned}$$

characterizing $(\tilde{z}^1, \tilde{z}^2, \tilde{\phi}^1, \tilde{\phi}^2)$ as a minimizer of $\mathcal{G}_\alpha$. □

4.2. Optimality conditions for the stabilized functional

Numerical algorithms for minimizing the stabilized functional (16) are typically based on attempts to satisfy the first order optimality conditions. To this end we consider $\mathcal{G}_{\varepsilon,\alpha}$ with $Y = L^2(\Gamma_2)$ and derive the directional derivatives with respect to ϕ^j for $j = 1, 2$, which read

$$R^j_{\varepsilon,\alpha,\beta} = 0, \qquad j = 1, 2 \tag{17}$$

with

$$\begin{aligned}
R^j_{\varepsilon,\alpha,\beta}(\phi^1, \phi^2) := {}& \Psi^j_\varepsilon H'_\varepsilon(\phi^j) L^*(L(P_\varepsilon(\phi^j, \phi^j)) - (g_2^\delta - z^\delta)) \\
& - \alpha \left[\frac{\beta}{2} \nabla \cdot \left(H'_\varepsilon(\phi^j) \nabla H_\varepsilon(\phi^j) / |\nabla H_\varepsilon(\phi^j)|\right) - (I - \Delta)(\phi^j - \phi^j_0)\right]
\end{aligned}$$

and

$$\Psi^1_\varepsilon(\phi^1, \phi^2) = (c_1 - c_2 - c_3 + c_4) H_\varepsilon(\phi^2) + (c_3 - c_4), \tag{18}$$

$$\Psi^2_\varepsilon(\phi^1, \phi^2) = (c_1 - c_2 - c_3 + c_4) H_\varepsilon(\phi^1) + (c_3 - c_4). \tag{19}$$

Below we consider certain fixed point iterations for solving the system of first order optimality conditions. Let us first note that (17) is equivalent to

$$M\begin{pmatrix}\phi^1-\phi^1_0\\ \phi^2-\phi^2_0\end{pmatrix} = M\begin{pmatrix}\phi^1-\phi^1_0\\ \phi^2-\phi^2_0\end{pmatrix} - \begin{pmatrix}R^1_{\varepsilon,\alpha,\beta}\\ R^2_{\varepsilon,\alpha,\beta}\end{pmatrix}. \tag{20}$$

If we let M vary in each iteration, we obtain algorithms of the following form

$$M_k\begin{pmatrix}\phi^1_{k+1}-\phi^1_k\\ \phi^2_{k+1}-\phi^1_k\end{pmatrix} = M_k\begin{pmatrix}\phi^1_k-\phi^1_0\\ \phi^2_k-\phi^2_0\end{pmatrix} - \begin{pmatrix}R^1_{\varepsilon,\alpha,\beta}(\phi^1_k,\phi^2_k)\\ R^2_{\varepsilon,\alpha,\beta}(\phi^1_k,\phi^2_k)\end{pmatrix}. \tag{21}$$

These iterations can be considered as preconditioned fixed-point iterations for (17).

4.3. Multiple level-set algorithms

In the following, we will consider three different choices for M_k.

Algorithm 1 (Simple iteration)**.** As a first method, we consider the fixed point iteration discussed in [15], which has the form (21) with

$$M_k := \begin{pmatrix}\alpha(I-\Delta) & 0\\ 0 & \alpha(I-\Delta)\end{pmatrix}.$$

In this case, the two equations in (21) decouple, and the iteration can be written as

$$\begin{aligned}\alpha(I-\Delta)(\phi^j_{k+1}-\phi^j_k) = {} & \Psi^j_\varepsilon H'_\varepsilon(\phi^j)L^*(L(P_\varepsilon(\phi^j,\phi^j)) - (g^\delta_2 - z^\delta))\\ & -\alpha\frac{\beta}{2}\nabla\cdot\left(H'_\varepsilon(\phi^j)\nabla H_\varepsilon(\phi^j)/|\nabla H_\varepsilon(\phi^j)|\right).\end{aligned}$$

Identifying $\alpha = 1/\Delta t$, $t_n = n\Delta t$, and $\phi^j_n = \phi^j(t_n)$, $n = 1, 2, \ldots$, we find that

$$(\Delta - I)\left(\frac{\phi^j(t_n)-\phi^j(t_{n-1})}{\Delta t}\right) = R^{j,*}_{\varepsilon,1/\Delta t,\beta}(\phi^1(t_{n-1}),\phi^2(t_{n-1})). \tag{22}$$

If we consider Δt as step length in a time discretization, we find that in a formal sense the iterative regularized solution $\phi^j_n, j = 1, 2$, is an approximate solution of the dynamical system

$$(\Delta - I)\left(\frac{\partial\phi^j(t)}{\partial t}\right) = R^j_{\varepsilon,1/\Delta t,\beta}(\phi^1(t),\phi^2(t)),\ j = 1,2. \tag{23}$$

Algorithm 2 (A Gauss-Newton algorithm)**.** Our second choice for M_k is motivated by Gauß-Newton methods for solving (17), i.e., we set

$$M_k := \begin{pmatrix}\psi^1_k H'_\varepsilon(\phi^1_k)L^*LH'_\varepsilon(\phi^1_k)\psi^1_k + \alpha(I-\Delta) & \psi^1_k H'_\varepsilon(\phi^1_k)L^*LH'_\varepsilon(\phi^2_k)\psi^2_k\\ \psi^2_k H'_\varepsilon(\phi^2_k)L^*LH'_\varepsilon(\phi^1_k)\psi^1_k & \psi^2_k H'_\varepsilon(\phi^2_k)L^*LH'_\varepsilon(\phi^2_k)\psi^2_k + \alpha(I-\Delta)\end{pmatrix}.$$

Here, M_k is just the Gauß-Newton approximation for the second derivative of the functional (16) with $\beta = 0$. While we expect faster convergence of the Newton-type method in comparison to the simple iteration, now a coupled linear system has to be solved in each step of iteration (21). Since M_k is positive definite, these linear systems can be solved iteratively, e.g., by a conjugate gradient method.

Journal of Physics: Conference Series **255** (2010) 012005 doi:10.1088/1742-6596/255/1/012005

Algorithm 3 (Preconditioned iterations). As a third alternative we consider the choice

$$M_k = \begin{pmatrix} (I-\Delta)^{-1} & 0 \\ 0 & (I-\Delta)^{-1} \end{pmatrix} M_k^*,$$

where M_k^* is either the matrix of the simple iteration or of the Gauß-Newton method above. For a detailed analysis of similar preconditioned iterative methods, we refer to [16, 17].

5. Numerial experiments

In this section we illustrate the advantages of the multiple levelset methods in reconstructing piecewise constant parameters over standard methods based on the regularized solution of the quadratice least-squares problem. Moreover, we compare the three different numerical methods for minimizing the minimizing the stabilized functional (16) outlined in the previous section.

5.1. The model problem and its discretization

For our numerical experiments we consider the following three dimensional Cauchy problem: Let $\alpha > 0$ be given, and consider the domain $\Omega := (0,1) \times (0,1) \times (0,a)$ with boundary $\partial\Omega = \overline{\Gamma_0} \cup \Gamma_a \cup \Gamma_L$ consisting of three parts

$$\Gamma_0 := (0,1)^2 \times \{0\}, \qquad \Gamma_a := (0,1)^2 \times \{a\}, \qquad \text{and} \qquad \Gamma_L := \partial\Omega \setminus \overline{\Gamma_0 \cup \Gamma_a}.$$

We consider the boundary value problem

$$-\Delta u = 0 \qquad \text{in } \Omega, \tag{24}$$

$$u = 0 \qquad \text{on } \Gamma_0 \cup \Gamma_L \tag{25}$$

$$u_\nu = \varphi \qquad \text{on } \Gamma_a, \tag{26}$$

and the corresponding Cauchy problem of determining the function ϕ from additional observations

$$u_\nu = g \qquad \text{on } \Gamma_0.$$

For the numerical solution of (24)–(26) we use a discretization method based on Fourier series. Let $\varphi_{n,m}$ denote the Fourier coefficients of ϕ, i.e.,

$$\varphi(x,y) = \sum_{m,n} \varphi_{m,n} \sin(m\pi x)\sin(n\pi y),$$

then the solution of the boundary value problem (24)–(26) is given by

$$u(x,y,z) = \sum_{m,n} u_{m,n} \sin(m\pi x)\sin(n\pi y)\sinh(\omega_{m,n}\pi a), \qquad \omega_{m,n} := \sqrt{m^2+n^2}$$

with coefficient $u_{m,n}$ defined by

$$u_{m,n} = \frac{\varphi_{m,n}}{\omega_{m,n}\pi\cosh(\omega_{m,n}\pi a)}.$$

The corresponding forward operator L in (6) is then given by

$$(L\varphi)(x,y) = \sum_{m,n} u_{m,n} \sin(m\pi x)\sin(n\pi y)\sinh(\omega_{m,n}\pi a).$$

5.2. Numerical tests

In our numerical experiments, we try to identify the piecewise constant function

$$\phi(x,y) = \begin{cases} 2, & (x-0.75)^2 + (y-0.75)^2 < 0.15 \\ 1, & (x-0.75)^2 + (y-0.25)^2 < 0.15 \text{ or } (x,y) \in (0.15, 0.35)^2 \\ 0, & \text{else} \end{cases}$$

from additional measurements of the Neumann trace $g := u_\nu$ at Γ_0. Throughout our numerical experiments we use synthetic data which are perturbed by random noise. The true solution and the corresponding data are displayed in Figure 1. In all numerical tests, we stop the iterations according to a discrepancy principle, i.e., we terminate the iterations when the first time the norm of the residual is less than $\tau\delta$, and we use $\tau = 1.5$ in our simulations. For the Gauß-Newton methods, we further utilize an adaptive strategy for choosing the regularization parameter α, i.e., we start with $\alpha = 1$ and decrease α in every Newton iteration by a factor 0.3. For the simple iterations, we choose $\alpha = \delta^2$. For the levelset methods, we utilize a

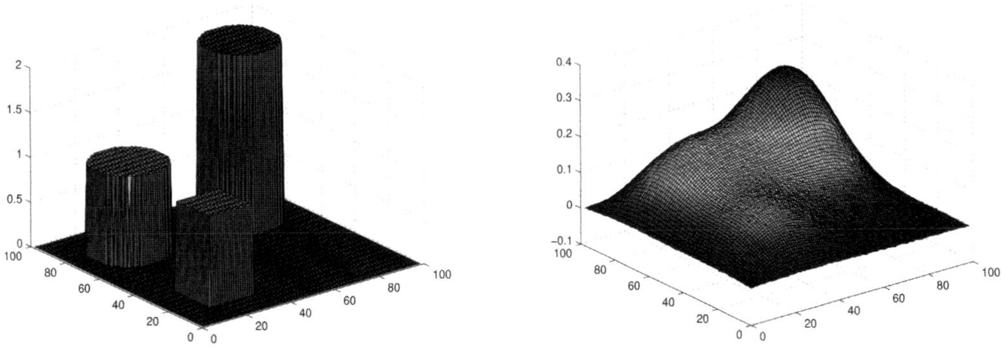

Figure 1. Exact solution (left) and data (right) perturbed with $\delta = 10^{-3}$ noise.

smoothed Heaviside projector

$$H_\varepsilon(z) := \frac{1}{2}(\text{erf}(z) + 1),$$

and we report only numerical experiments with $\beta = 0$, i.e., without bounded variation regularization term. Since the true solution only attains three different values, we choose $c_1 = 0$, $c_2 = 1$, $c_2 = 2$ and $c_4 = 0$ in the definition of the "projection" operator P, see (11).

In Figure 2, we display the solutions obtained with a standard conjugate gradient method applied to the linear Cauchy problem (6), and the one obtained with the preconditioned Gauß-Newton levelset method outlined in Algorithm 2 and 3. The reconstructions obtained with the different levelset algorithms are very similar, and we therefore only present the results for one of the methods. Also the reconstruction errors of the the methods are comparable, and we only list the error of conjugate gradient method applied to the solution of the linear Cauchy problem and the preconditioned Gauß-Newton method in Table 1. The re reconstructions of the levelset methods are clearly superior to the ones obtained with standard regularization

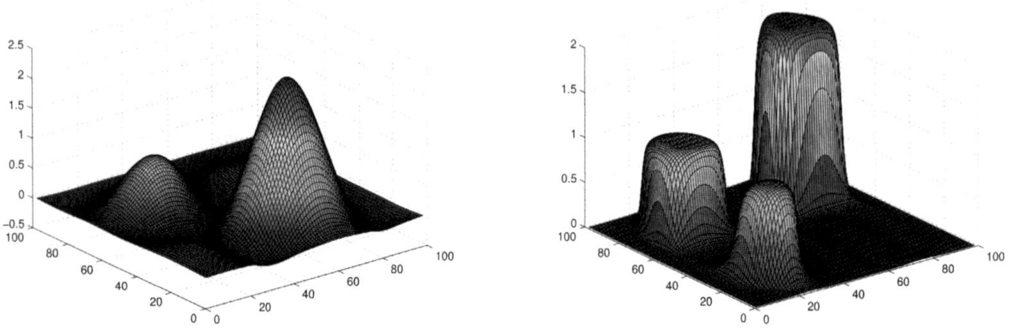

Figure 2. Comparison of reconstructions obtained with conjugate gradients method applied to the linear Cauchy problem (left), and the preconditioned Gauß-Newton levelset method (right) for noise level $\delta = 10^{-3}$.

δ	0.01	0.001	0.0001	0.00001
CGNE	0.3940	0.2615	0.2356	0.2185
P-GN	0.2256	0.1804	0.1227	0.0763

Table 1. Reconstruction errors $\|\varphi^\dagger - \varphi_k\|_0$ of CGNE applied to the linear Cauchy problem (6), and the preconditioned Gauß-Newton levelset method (Algorithm 3) for varying noise level; see also Figure 2.

methods, e.g., the L^2 reconstruction error of the levelset methods at noise level $\delta = 10^{-2}$ is comparable to the one of the CGNE method with noise level $\delta = 10^{-4}$. This shows that the utilization of a-priori knowledge, i.e. the assumption of a piecewise constant solution, in the formulation of the algorithm can drastically improve the quality of reconstructions.

While the solutions obtained with the different levelset algorithms are very similar, the computational cost of Algorithms 1–3 varies significantly. In Table 2, we compare the iteration numbers of the simple iteration and the Gauß-Newton algorithm, and their preconditioned variants.

δ	SIM	GN	P-SIM	P-GN
0.01	30	7 (10)	9	3 (10)
0.001	575	13 (37)	70	6 (28)
0.0001	>10000	20 (113)	4272	9 (62)

Table 2. Iteration numbers for the simple iteration (SIM) and the Gauß-Newton method (GN), as well as their preconditioned variants for various noise levels. For the Gauß-Newton iterations, the total number of inner CG iterations are listed in parenthesis.

Acknowledgments
The authors would like to thank Dr. H. Egger (MathCCES, Aachen) for valuable discussions and for performing the numerical computations. The work of A.L. is partially supported by CNPq, grant 303098/2009-0 and by the Alexander von Humboldt Foundation AvH.

References
[1] R.A. Adams, *Sobolev Spaces,* Academic Press, New York, 1975
[2] G. Alessandrini and E. Sincich, *Solving elliptic Cauchy problems and the identification of nonlinear corrosion,* J. Comput. Appl. Math. **198** (2007), 307–320
[3] S. Andrieux, T.N. Baranger and A. Ben Abda, *Solving Cauchy problems by minimizing an energy-like functional,* Inverse Problems 22 (2006), 115–133
[4] G. Aubert and P. Kornprobst, *M*athematical Problems in Image Processing. Partial Differential Equations and Calculus of Variations, Springer, New York, 2006
[5] F.B. Belgacem, *Why is the Cauchy problem severely ill-posed?* Inverse Problems 23 (2007), 823–836
[6] L. Bourgeois, *A mixed formulation of quasi-reversibility to solve the Cauchy problem for Laplace's equation,* Inverse Problems 21 (2005), 1087–1104
[7] L. Bourgeois, *Convergence rates for the quasi-reversibility method to solve the Cauchy problem for Laplace's equation,* Inverse Problems 22 (2006), 413–430
[8] M. Burger, *A level set method for inverse problems,* Inverse Problems 17 (2001), 1327–1355
[9] F. Cakoni and R. Kress, *Integral equations for inverse problems in corrosion detection from partial Cauchy data,* Inverse Problems and Imaging 1 (2007), 229–245
[10] A.-P. Calderón, *Uniqueness in the Cauchy problem for partial differential equations,* Amer. J. Math. 80 (1958), 16–36
[11] H. Cao and S.V. Pereverzev, *The balancing principle for the regularization of elliptic Cauchy problems,* Inverse Problems **23** (2007), 1943–1962
[12] A. Chakib and A. Nachaoui, *Convergence analysis for finite element method approximation to an inverse Cauchy problem,* Inverse Problems 22 (2006), 1191–1206
[13] T.F. Chan and X.C. Tai, *Level set and total variation regularization for elliptic inverse problems with discontinuous coefficients*, J. Comput. Phys.193 (2004), 40–66
[14] R. Dautray and J.L. Lions, *Mathematical Analysis and Numerical Methods for Science and Technology,* Vol. 2, Springer, New York, 1988
[15] A. De Cezaro, A. Leitão, and X.-C. Tai, *On multiple level-set regularization methods for inverse problems,* Inverse Problems 25 (2009), 035004
[16] H. Egger and A. Neubauer, *Preconditioning Landweber iteration in Hilbert scales*, Numer. Math. 101 (2005), 643-662
[17] H. Egger, *Fast fully iterative Newton-type methods for inverse problems,* J. Inv. Ill-posed Problems 15 (2007), 257–276
[18] H. Egger and A. Leitão, *Efficient reconstruction methods for nonlinear elliptic Cauchy problems with piecewise constant solutions,* Advances in Applied Mathematics and Mechanics 1 (2009), 729–749
[19] H. Engl, M. Hanke and A. Neubauer, *Regularization of Inverse Problems,* Kluwer, Dordrecht, 1996
[20] H. Engl and A. Leitão, *A Mann iterative regularization method for elliptic Cauchy problems,* Numer. Funct. Anal. Optim. 22 (2001), 861–864
[21] L.C. Evans and R.F. Gariepy, *Measure Theory and Fine Properties of Functions,* CRC Press, Boca Raton, 1992
[22] R.S. Falk and P.B. Monk, *Logarithmic convexity of discrete harmonic functions and the approximation of the Cauchy problem for Poisson's equation*, Math. Comput. 47 (1986), 135-149
[23] F. Frühauf, O. Scherzer and A. Leitão, *Analysis of regularization methods for the solution of ill-posed problems involving discontinuous operators*, SIAM Journal of Numerical Analysis 43 (2005), 767–786
[24] D. Gilbarg and N.S. Trudinger, *Elliptic partial differential equations of second order,* Springer, Berlin, 1977
[25] J. Hadamard, *Le problème de Cauchy et les équations aux dérivées partielles linéaires hyperboliques,* Hermann, Paris, 1932
[26] D.N. Hào and D. Lesnic, *The Cauchy problem for Laplace's equation via the conjugate gradient method* IMA J. Appl. Math. 65 (2000), 199–217

[27] Y.C. Hon and T. Wei, *Backus-Gilbert algorithm for the Cauchy problem of the Laplace equation,* Inverse Problems 17 (2001), 261–271

[28] G. Inglese, *An inverse problem in corrosion detection,* Inverse Problems 13 (1997), 977–994

[29] V.A. Kozlov, V.G. May'za and A.V. Fomin, *An iterative method for solving the Cauchy problem for elliptic equations,* Comput. Maths. Phys. 1 (1991), 45–52

[30] P. Kügler and A. Leitão, *Mean value iterations for nonlinear elliptic Cauchy problems,* Numerische Mathematik 96 (2003), 269–293

[31] A. Leitão, *Applications of the Backus–Gilbert method to linear and some non linear equations,* Inverse Problems 14 (1998), 1285–1297

[32] A. Leitão, *An iterative method for solving elliptic Cauchy problems,* Numer. Funct. Anal. Optim. 21 (2000), 715–742

[33] A. Leitão and M. Marques Alves, *On level set type methods for elliptic Cauchy problems,* Inverse Problems 23 (2007), 2207–2222

[34] A. Leitão and O. Scherzer, *On the relation between constraint regularization, level sets, and shape optimization,* Inverse Problems 19 (2003), L1–L11

[35] J.-L. Lions, *Optimal control of systems governed by partial differential equations,* Springer, New York, 1971

[36] F. Santosa, *A level set approach for inverse problems involving obstacles,* ESAIM: Control, optimization and Calculus of Variations 1 (1996), 17–33

[37] R.E. Showalter, *Monotone Operators in Banach Space and Nonlinear Partial Differential Equations,* American Mathematical Society, Providence, 1997

[38] U. Tautenhahn, *On the asymptotical regularization of nonlinear ill-posed problems,* Inverse Problems 10 (1994), 1405–1418

[39] K. van den Doel, U.M. Ascher, *D*ynamic level set regularization for large distributed parameter estimation problems, Inverse Problems **23** (2007), 1271–1288

Development of in-vivo fluorescence imaging with the Matrix-Free method

Athanasios Zacharopoulos†, Anikitos Garofalakis §, Jorge Ripoll †, Simon Arridge ‡

†Institute for Electronic Structure and Laser, Foundation for Reasearch and Technology-Hellas, P.O. Box 1527, Heraklion 71110, Greece, §CEA, I2BM, Service Hospitalier Frèdèric Joliot, Laboratoire d'Imagerie Molèculaire Expèrimentale, 4 place du gènèral Leclerc, 91401 Orsay Cedex, France ‡Dept. of Computer Science, UCL,Gower st, WC1E 6BT, UK

E-mail: azacharo@iesl.forth.gr

Abstract. Non-contact Fluorescence Molecular Tomography is an emerging technique for imaging of fluorescent probes or proteins in live animals. One of the main characteristics of the non-contact acquisition systems in comparison to the usual fibre-based systems is the much denser boundary data sets that are created. When model-based reconstruction methods are used that rely on the inversion of a derivative operator, the large number of measurements poses a challenge since the explicit formulation and storage of the Jacobian matrix could be in general not feasible. In this paper we test a matrix-free method that addresses the problems of large data sets and reduces the computational cost and memory requirements for the reconstruction. More specifically we challenged the Matrix-Free method with in-vivo measurements from mice where fluorescence tubes of different but controlled concentrations are inserted, to assess the quantification performance of the method. We extended the test with simulations, using realistic geometries extracted from a mouse-atlas and including prior known information about the optical properties of tissue into the forward model.

1. Introduction

Recent advances in optical technologies and the development of several fluorescent proteins and switchable fluorochromes have evolved the field of optical imaging within the frame of Molecular Imaging, which is constantly gaining importance in Life Sciences. Compounds exhibiting fluorescent properties have recently played a major role in many biotechnology applications such as gene expression profiling, determining protein function as well as in analyzing cellular pathways and small-molecule/protein-protein interactions. Tomographic approaches have been utilized to image novel fluorescent agents with functional and molecular specificity through several millimeters to centimeters of tissue in vivo [1].

It is common for the developed fluorescence tomographic systems [2, 4, 3] to employ non contact detection schemes in order to retrieve the fluorescence measurements by means of an optical system that projects the surface of the medium onto a CCD camera producing a large amount of measurements from the pixels of the CCD sensor. Provided that the signal-to-noise ratio is good enough, a larger dataset can provide an improved solution to the inverse problem; firstly, by the increased spatial resolution of the reconstruction [5] and secondly,by reducing the illposedness of the problem by the higher information content [6].

In order to solve the inverse tomographic problem with a large number of data the requirements on the computational hardware become immense. In the most common reconstruction schemes the system Jacobian is built and stored in the computer memory and the size of the Jacobian matrix depends on the number of measurements acquired and the resolution of the geometry used for the numerical solution. Practically this imposes a limit for the maximum size of the Jacobian depending on the amount of computer memory that is available.

The problem of large datasets has been investigated for the diffuse optical tomography case using an approach based on the analytical solution to the diffusion equation for geometries where the solution is available in [8] and [7]. They report on the ability to reconstruct absorption heterogeneities using a number of source-detector pairs in the order of 10^8.

Similarly, in [9] a hybrid Fourier encoding for the measurements has been used to successfully reduce the effect of the large datasets for certain geometries. In addition, compression with wavelets has been implemented for reconstructions from simulations in [10] on realistic geometries.

Our approach initially presented in [11], is based on a formulation of the diffusion approximation for the fluorescence case and uses a matrix free formulation. Hence, the explicit calculation and storage of the Jacobian is avoided by replacing it by a vector times matrix operator and a vector times adjoint matrix operator. Also, the matrix free formulation has the property of requiring equal computer power independent on the number of detectors used, thus making it ideal for imaging with large number of measurements.

In this paper we initially present in-vivo reconstructions from mouse experiments with controlled concentrations of Alexa Fluor 647 (Invitrogen, France). For the reconstructions we have used a slab approximation for the mouse geometry and the matrix-free method. Having five different sets of measurements with different concentrations of the fluorochrome, results are also presented on the quantification capabilities of the method.

Since fluorescence tomography is a functional imaging modality in the recent years the inclusion of prior geometric information in the inversion of fluorescence is becoming a new common practice. Anatomical prior information of the biological tissue is usually acquired from a second modality on site, like X-ray Computed Tomography (XCT) [16, 15] or Magnetic Resonance Imaging (MRI) [14] to increase imaging resolution. A feasibility study using 2D simulations has shown the importance of a-priori structural and functional information (optical parameters) for the reconstruction of fluorescence targets in the brain [23]. To this end, in addition to the in-vivo experiments we performed a simulation study, where anatomical information from a mouse atlas was incorporated in the forward problem and a fast matrix-free solution of the inverse problem is presented.

This paper is organized as follows. In subsection 2.1 we describe the problem using a set of two paired diffusion equations for the excitation and the fluorescence wavelength, and construct the derivative operators. Subsection 2.2 defines the reconstruction as an optimization problem and gives an insight to the scalings that were introduced to improve the convergence of the problem. Subsection 2.3 presents the matrix-free approach. In vivo reconstructions from mice and the quantification properties of the method is the subject of section 3, and finally reconstructions using simulations with realistic geometries and a-priori known optical parameters are presented in section 4.

2. Matrix Free Method

2.1. Forward Model

The forward model for the continuous-wave fluorescence tomography problem is given by the coupled diffusion equations at the excitation and emission wavelengths λ_e and λ_f, respectively:

$$\begin{aligned}(-\nabla D(\mathbf{r},\lambda_e)\nabla + \mu_{\mathrm{a}}(\mathbf{r},\lambda_e))\, U^{(e)}(\mathbf{r}) &= q(\mathbf{r}) &(1)\\ (-\nabla D(\mathbf{r},\lambda_f)\nabla + \mu_{\mathrm{a}}(\mathbf{r},\lambda_f))\, U^{(f)}(\mathbf{r}) &= U^{(e)}(\mathbf{r})h(\mathbf{r},\lambda_f), &(2)\end{aligned}$$

with Robin boundary condition at both wavelengths, [12],

$$U(\xi) + 2\zeta D(\xi,\lambda)\frac{\partial U^{(e/f)}(\xi)}{\partial n} = 0, \qquad \xi \in \partial\Omega \tag{3}$$

where D and μ_{a} are the wavelength-dependent diffusion and absorption coefficients, q is a boundary source at the excitation wavelength, ζ is a boundary term incorporating the refractive index mismatch at the surface of the medium, n is the outward surface normal at ξ, and $U^{(e)}$ and $U^{(f)}$ are the photon density fields at the excitation and emission wavelength, respectively. We have defined the wavelength-dependent *fluorescence yield coefficient* h for the domain Ω:

$$h(\mathbf{r},\lambda) = \eta(\lambda)c(\mathbf{r}), \qquad \mathbf{r} \in \Omega$$

where $\eta(\lambda) = \epsilon\gamma(\lambda)$ is the product of the quantum yield $\gamma(\lambda)$ and the extinction coefficient ϵ for the fluorochrome at wavelength λ, and $c(\mathbf{r})$ is the concentration distribution of the fluorochrome. The contribution of h to the absorption at λ_e is here considered negligible.

The exitance at boundary $\partial\Omega$ for both wavelengths is given by the boundary operator

$$y(\xi) = -D(\xi,\lambda)\frac{\partial U(\xi)}{\partial n} = \frac{1}{2\zeta}U(\xi) \qquad \xi \in \partial\Omega \tag{4}$$

The exitance distribution on the surface is sampled with an array of detectors (e.g. a CCD camera), providing a discrete set of measurements $g_d^{(e)}$ and $g_d^{(f)}$ at both wavelengths, such that

$$g_d^{(e/f)} = \mathcal{M}\left[y^{(e/f)}\right] := \int_{\partial\Omega} y^{(e/f)}(\xi)w_d(\xi)d\xi \tag{5}$$

where w_d is the sensitivity profile of detector d on the boundary.

Assuming that the optical parameters $D(\mathbf{r},\lambda)$ and $\mu_{\mathrm{a}}(\mathbf{r},\lambda)$ are known, the reconstruction problem consists in finding the fluorochrome concentration $c(\mathbf{r})$. First consider the problem of reconstructing h. Eq. (1-5) define the forward problem which maps distribution $h(\mathbf{r},\lambda_f)$ to fluorescence data $g_d^{(f)}$:

$$\mathbf{g}^{(f)} = F(h) \tag{6}$$

To solve Eq. (1) and (2), we use a finite element model. Parameters D, μ_{a} and h are expressed as finite-dimensional vectors $\mathbf{D}$, μ_{a}, $\mathbf{h} \in \mathbb{R}^P$ whose elements are the coefficients of a basis expansion

$$D(\mathbf{r}) \approx \sum_{k=1}^{P} D_k b_k(\mathbf{r}), \qquad \mu_{\mathrm{a}}(\mathbf{r}) \approx \sum_{k=1}^{P} \mu_k b_k(\mathbf{r}), \qquad h(\mathbf{r}) \approx \sum_{k=1}^{P} h_k b_k(\mathbf{r}) \tag{7}$$

with basis functions $\mathbb{B} = \{b_k(\mathbf{r}), k = 1 \ldots P\}$. The fields U are expanded in the basis of a finite element methods (FEM) :

$$U(\mathbf{r}) \approx \sum_{\ell=1}^{N} U_\ell v_k(\mathbf{r}) \tag{8}$$

hence the diffusion equation can be written as a linear system

$$\mathsf{K}(\mathbf{D},\mu)\mathbf{U} = \mathbf{Q} \tag{9}$$

where $\mathsf{K} \in \mathbb{R}^{N\times N}$ is a system matrix assembled from element contributions that depend on the parameter distributions, and $\mathbf{U}$ is the vector of basis coefficients representing the photon density distribution in basis expansion $\mathbb{V} = \{v_\ell(\mathbf{r}), \ell = 1 \dots N\}$. The forward model in the discrete setting consists of solving

$$\mathsf{K}_{\lambda_e}\mathbf{U}^{(e)} = \mathbf{Q}, \tag{10}$$

$$\mathsf{K}_{\lambda_f}\mathbf{U}^{(f)} = \mathbf{h}\odot\mathbf{U}^{(e)}, \tag{11}$$

with $\mathsf{K}_{\lambda_f} = \mathsf{K}[\mathbf{D}(\lambda_f),\mu_{\mathrm{a}}(\lambda_f)]$, $\mathsf{K}_{\lambda_e} = \mathsf{K}[\mathbf{D}(\lambda_e),\mu_{\mathrm{a}}(\lambda_e)]$ and $\odot$ representing element-wise multiplication. Together with the boundary operator, the forward model can now be expressed in the linear form

$$\mathbf{g}_{\mathbf{Q}}^{(f)} = F_{\mathbf{Q}}(\mathbf{h}) = \mathsf{A}_{\mathbf{Q}}^{(h)}\mathbf{h} := \mathcal{M}\left[\mathsf{K}_{\lambda_f}^{-1}\mathbf{h}\mathsf{K}_{\lambda_e}^{-1}\mathbf{Q}\right] \tag{12}$$

where $\mathsf{A}^{(h)}$ is the discrete matrix representation of the forward operator. Examination of the expression in Eq. (12) indicates that the Jacobian can be constructed using the adjoint formulation

$$J_{(sd),k}^{(h)} = U_{s,k}^{(e)}U_{d,k}^{(f)+} \tag{13}$$

where (sd) denotes a row index constructed from an ordering of source index s and measurement index d, and

$$\mathsf{K}_{\lambda_e}\mathbf{U}_s^{(e)} = \mathbf{Q}_s, \tag{14}$$

$$\mathsf{K}_{\lambda_f}\mathbf{U}_d^{(f)+} = \mathbf{Q}_d^{+}, \tag{15}$$

are the solutions of the direct and adjoint problems.

2.2. Inverse problem

In general we pose the inverse problem as an optimisation problem

$$\hat{\mathbf{x}} = \arg\min_x\left[\Phi(\mathbf{x}) := \frac{1}{2}||\mathbf{g}_{\mathrm{meas}}^{(\mathrm{f})} - F(\mathbf{x})||^2 + \alpha\Psi(\mathbf{x})\right] \tag{16}$$

with regularisation term Ψ and hyperparameter α. Since $F(\mathbf{x}) = \mathsf{A}\mathbf{x}$ is linear, an iterative solution to Eq. (16) is obtained for example by the damped Gauss-Newton scheme :

$$\begin{aligned}\left(\mathsf{A}^{\mathrm{T}}\mathsf{A} + \alpha\Psi''(\mathbf{x}^{(k)})\right)\mathbf{x}^{\delta} &= \mathsf{A}^{\mathrm{T}}(\mathbf{g}_{\mathrm{meas}}^{(\mathrm{f})} - \mathsf{A}\mathbf{x}^{(k)}) - \alpha\Psi'(\mathbf{x}^{(k)}) \\ \tau_k &= \arg\min_{\tau}\Phi(\mathbf{x}^{(k)} + \tau\mathbf{x}^{\delta}) \\ \mathbf{x}^{(k+1)} &= \mathbf{x}^{(k)} + \tau_k\mathbf{x}^{\delta}\end{aligned} \tag{17}$$

where τ denotes the step length. If the regularisation is quadratic, $\Psi(\mathbf{x}) = \frac{1}{2}||\mathsf{L}\mathbf{x}||^2$, we have the direct reconstruction formula

$$(\mathsf{A}^{\mathrm{T}}\mathsf{A} + \alpha\mathsf{L}^{\mathrm{T}}\mathsf{L})\hat{\mathbf{x}} = \mathsf{A}^{\mathrm{T}}\mathbf{g}_{\mathrm{meas}}^{(\mathrm{f})} \tag{18}$$

where L is for example the Cholesky factorisation of a quadratic Markov Random field. In the results presented in this paper, we consider only the simplest *zero-order Tikhonov* regularisation $\mathsf{L} = \mathsf{I}$.

Due to uncertainties in laser power, detector gain, and losses we cannot expect the forward model to be compatible with the measured data. To avoid problems due to the model mismatch, we employ data normalization by making use of the available excitation data, and thus we do not require absolute measurements. The optimization problem Eq. (16) is transformed into the rescaled problem

$$\hat{\mathbf{x}} = \arg\min_{\mathbf{x}} \frac{1}{2}||\tilde{\mathbf{g}}^{(f)} - \tilde{F}(\mathbf{x})||^2 + \alpha\Psi(\mathbf{x}) \tag{19}$$

We used the Normalized Born approximation [20], which is extensively used in fluorescence tomography.

$$\mathbf{g}^{(\mathrm{f})}_{\mathrm{meas}} \rightarrow \tilde{\mathbf{g}}^{(f)} = \mathbf{g}^{(\mathrm{f})}_{\mathrm{meas}}, \qquad F(x) \rightarrow \tilde{F}(x) = \frac{\mathbf{g}^{(\mathrm{e})}_{\mathrm{meas}}}{\mathbf{g}^{(\mathrm{e})}_{\mathrm{proj}}} F(x) \tag{20}$$

Where $\mathbf{g}^{(\mathrm{e})}_{\mathrm{proj}}$ denotes the calculated data in the excitation wavelength.

2.3. Implementation of the matrix-free method

When faced with solving the linear problem Eq. (19), the explicit computation and storage of the matrices $\mathsf{A}^{\mathrm{T}}\mathsf{A}$ and $\mathsf{A}^{\mathrm{T}}\mathbf{g}^{(\mathrm{f})}_{\mathrm{meas}}$ is costly and often intractable for large scale problems. This is mainly due to the large size of the Jacobian matrix A. As an example, a problem with 30 sources and 475 detectors solved on a geometry with 7812 elements would need about 890MB of storage for each wavelength. The problem is more intense when more samples from the CCD image are required or a high resolution mesh is used to represent the domain in the solver. When using a Krylov solver for the linear problem we require to construct the set of basis vectors

$$\left\{\mathbf{z}, \mathsf{H}\mathbf{z}, \ldots \mathsf{H}^{j}\mathbf{z}\right\} \tag{21}$$

where $\mathbf{z} = \mathsf{A}^{\mathrm{T}}\mathbf{g}^{(\mathrm{f})}_{\mathrm{meas}}$ and $\mathsf{H} = \mathsf{A}^{\mathrm{T}}\mathsf{A} + \alpha\mathsf{L}^{\mathrm{T}}\mathsf{L}$. In our approach therefore, we represent the forward and adjoint multiplication by the Jacobian implicitly, using a function that returns the result of the operation. To be specific, we use the Matlab 'gmres' function to solve Eq. (19), and pass it a function that calculates the forward and adjoint solutions for any intermediate vector generated in the solution iteration. The method is summarised in Algorithm 1.

Algorithm 1 Schematic of reconstruction method

For each source s, calculate forward excitation fields $\mathbf{U}_s^{(e)}$
Calculate $\mathbf{z} = \mathsf{A}^{\mathrm{T}}\mathbf{g}^{(\mathrm{f})}_{\mathrm{meas}}$ using adjoint solver
Inside GMRES solver :
  $\mathbf{v}_0 = \mathbf{z}$.
  for all Krylov basis vectors $\mathbf{v}_j, j = 1\ldots$ **do**
    Update fluorphore concentration $\mathbf{h} = \sum_i \eta_i(\lambda_f)\mathbf{v}_{j-1,i}$
    Calculate forward projection $\mathbf{r} = \mathsf{A}\mathbf{h}$
      1 **Solve for every source** s **:** $\mathsf{K}_{\lambda_f}\mathbf{U}_s^{(\lambda_f)} = \mathbf{h} \odot \mathbf{U}_s^{(e)}$
      2 **Calculate :** $\mathbf{r}_s^f = \mathcal{M}[\mathbf{U}_s^{(\lambda_f)}]$
     Adjoint Calculation $\mathbf{x}^f = \mathsf{A}^{\mathrm{T}} \cdot \mathbf{r}^f$
    1 **Solve for every source** s **:** $\mathsf{K}_{\lambda_f}\mathbf{V} = \mathbf{r}_s^{(f)}$
    2 $\mathbf{f} = \mathbf{f} + \mathbf{V} \odot \mathbf{U}_s^{(e)}$
  end for

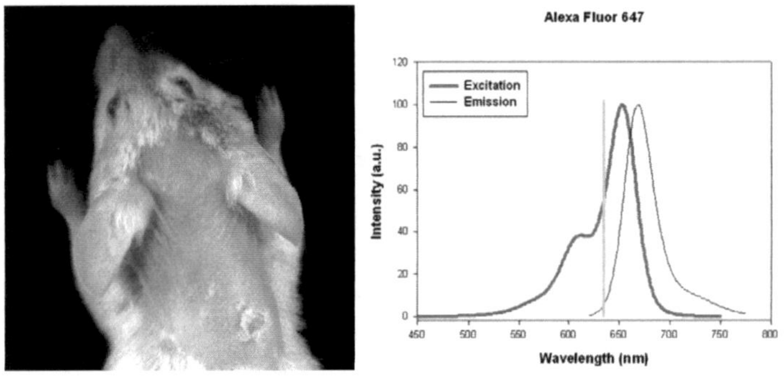

Figure 1. (left) The whitelight image of the mouse used for the Alexa Fluor capillary inserted in the ventral side of the mouse. (right) the spectral response for the Alexa Fluor 647 where we show for reference the laser line at 635nm.

3. In-vivo quantification experiments

In this section we assess the quantification properties of the matrix-free method using in-vivo measurements on mice. More specifically we used a balb/c mouse where capillary of Alexa Fluor 647 of controlled concentrations $0.9\mu M$(micromoles/litre) , to $60.5\mu M$ was inserted hypodermically in the ventral side of the mouse. The acquisition was performed using the experimental setup in FORTH, [21, 22]. The system incorporates a solid state laser at 635nm (Coherent, USA). The laser light is directed inside a light shielded imaging chamber and enters a laser scanning device (Scancube 7, Scanlab, Germany) which consists of two mirrors driven by miniaturized galvanometer motors. The angular movement of the motors is controlled by a custom made computer software for directing the laser beam in a controlled manner to the sample with the aid of a large rectangular mirror (first surface mirrors, 4-6 Wave, 73 mm 116 mm, Edmund Optics). The sample is placed on a transparent glass plate with an anti-reflection coating (Glassplate, High AR coated 96-99% (400-700nm)) and receives the illumination from below. Images are captured by a thermoelectrically cooled 16bit CCD camera with a chip size of 1024x1024 pixels (Andor Corp., DV434, Belfast, Northern Ireland), which is mounted on the upper plate of the imaging box, on the other side of the illumination. The CCD camera is equipped with a SIGMA 50mm f/2.8 objective (Sigma Corporation, Tokyo, Japan) which focuses on the sample's surface. A filter wheel mounted in front of the CCD camera lens enables the capturing of images at 2 wavelength positions. The first being at the emission range of the fluorophore, for taking the Fluorescence set of images and the second is at the excitation wavelength taking the Intrinsic set of data

Transmission measurements were acquired with the laser on the ventral side of the mouse and the camera facing the dorsal side, figure 1.The mouse was gently compressed by a transparent glass plate of 1mm thickness conforming the subject into a slab geometry. The measurements consisted of 36 sources position each collecting two 512x512 images, one for each wavelength. For the reconstructions, a slab-like geometry of dimensions 41mm x 41mm x 12mm was used with 5733 nodes and 4800 voxel-elements. For the optical parameters we assumed $\mu_a = 0.06mm^{-1}$ and $\mu_s' = 1.6mm^{-1}$ for the 615nm excitation and $\mu_a = 0.05mm^{-1}$ and $\mu_s' = 1.5mm^{-1}$ for the 700nm emission wavelengths. We set 1458 detectors positions by sampling the images, which for the 36 sources created two vectors of 48285 measurements each.

The results for the five different experiments are presented in figure 2, as slices that go through the center of the reconstructed capillaries in each case and represent a coronal plane slice along the y-axis (top) and a a transverse plane slice along the z-axis (bottom). The same results are also presented as iso-surfaces in 3D in figure 3, where the white light image of the mouse is

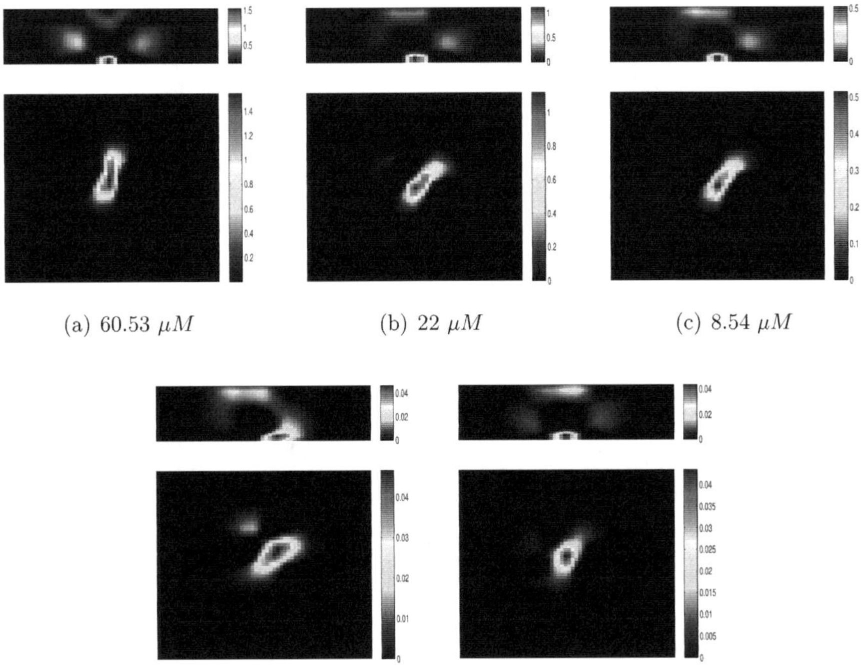

(a) 60.53 μM (b) 22 μM (c) 8.54 μM

(d) 2.23 μM (e) 0.91 μM

Figure 2. Reconstructions from an in-vivo mouse where capillaries filed with Alexa Fluor 650 of controlled concentrations have been inserted in the ventral side of the mouse, while the camera "looks" on the dorsal side. The different reconstructions correspond to concentrations a)60.5μM, b)22μM, c)8.5μM, d)2.2μM and e)0.9 μM. The slices go through the center of the reconstructed capillaries and represent a coronal plane slice along the y-axis (top) and a a transverse plane slice along the z-axis (bottom).

rendered in the background as a reference. The average time for the reconstruction was 50sec, in a Pentium 2Ghz machine with 2GB of memory.

Since, we had experiments with different concentrations of Alexa Fluor, we decided to assess the quantification accuracy of the matrix-free method. More specifically, we calculated a histogram of the reconstructed values for each experiment and took the mean value of the voxels with the highest concentrations. The results are presented in the figure 4, versus the experimentally known concentrations. The curve that connects the points on the graph, is almost linear, up to some experimental error, and shows the correspondence of the real concentrations to the reconstructed values.

4. Simulations with realistic geometries and prior known optical parameters

In the latest approaches in Fluorescence Imaging we see the inclusion of structural information for the mouse anatomy, either through the use of a second modality on-site, (XCT,MRI) or through the utilization of a mouse atlas, we decided to test the ability and performance of the Matrix-Free method on the use of realistic geometries. As an initial approach to the feasibility of the method with realistic geometries we decided to use an already existing atlas for the creation of simulated measurements and the reconstruction for a fluorescence target. More specifically we used geometries extracted from Digimouse, [17, 18], an adult mouse atlas generated using

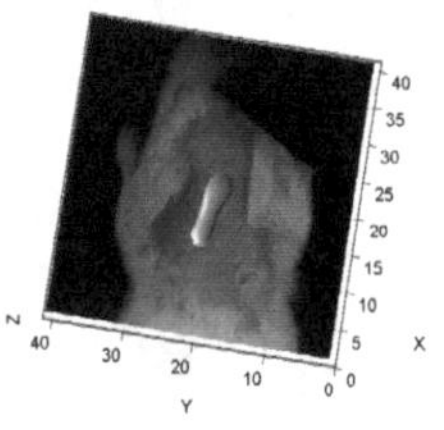

(a) 60.53 μM

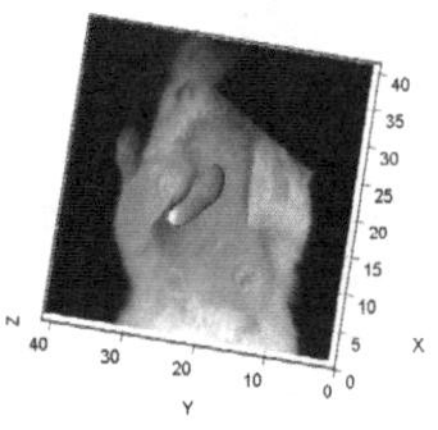

(b) 22 μM

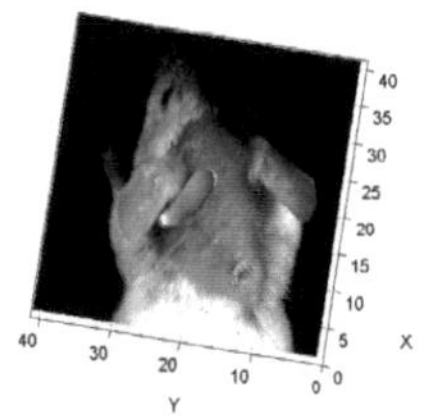

(c) 8.54 μM

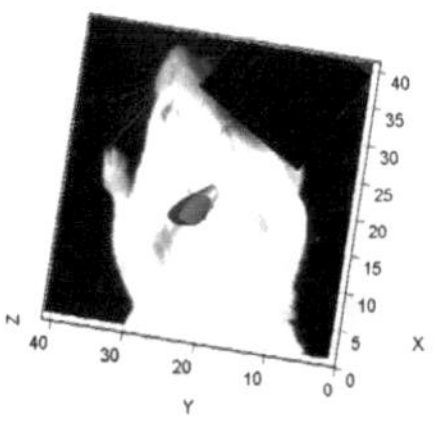

(d) 2.23 μM

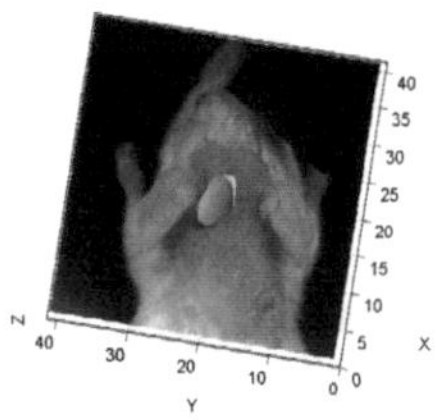

(e) 0.91 μM

Figure 3. Reconstructions from an in-vivo mouse where capillaries filed with Alexa Fluor 650 of controlled concentrations. 3D renderings of the reconstructed capillaries, for the different concentrations a)60.5μM, b)22μM, c)8.5μM, d)2.2μM and e)0.9 μM. The white light image of the mouse from the camera is rendered as a reference. The average time for the reconstruction was 50sec.

coregistered CT and cryosection images of a 28g nude normal male mouse. We extracted the surfaces of the head, brain and skull, from the atlas and using iso2mesh [19] mesh creation algorithm we created a mesh of 11714 nodes, and 60822 elements. An outline of the mesh with the three distinct regions can be seen in figure 5.

For the forward model the values for μ_{a} and μ_{s}' for the brain and the rest of the tissue were taken from [13] for the wavelengths of excitation at 682nm and emission at 710nm as seen in table 1. The fluorescence target had the shape of a rod and was placed inside the region of the brain as seen in figure 6 on the left side. We assumed a fluorescence response of 1.7 for the rod and zero for the rest of the tissue. 20 sources were placed in the ventral side of the mouse and 1584 detectors positions were assumed on the dorsal side. In total that produced 31680 measurements in each wavelength. Simulated data were created using the forward solution on the above setup. We added 5% multiplicative gaussian noise to the data and used them for the

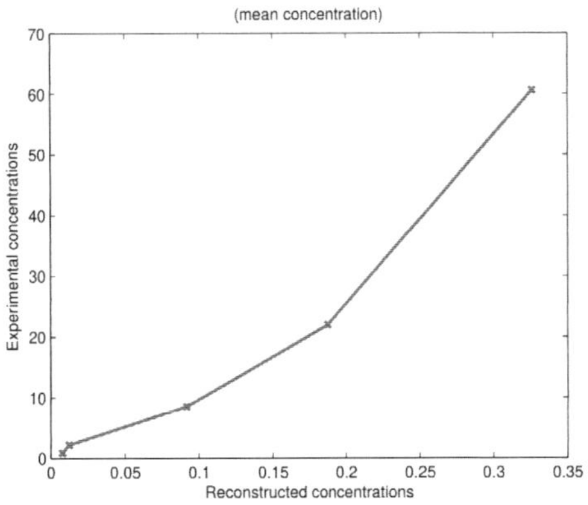

Figure 4. Plot of the reconstructed concentration for Alexa Fluor 647 from five experiments with different capillaries versus the experimentally known concentration.

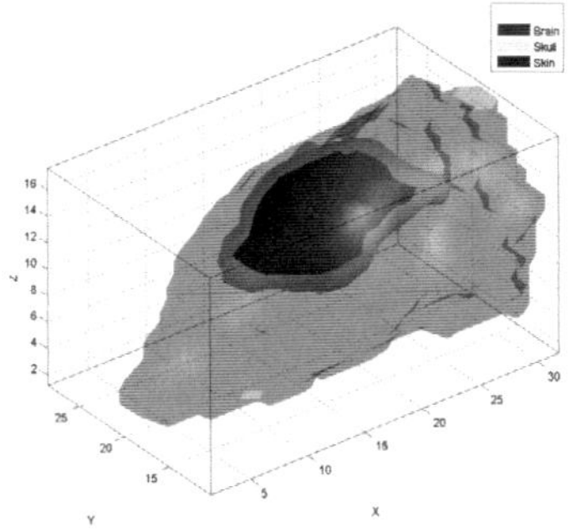

Figure 5. The geometrical model extracted from an atlas for the mouse head. The different regions correspond to different absorption and scattering parameters for the forward model.

Table 1. Optical parameters for the 682nm and the 710nm that were used for the simulation of the three regions in the mouse head.

Wavelength	Brain	Skull	Other tissue
682nm	μ_a= 0.023 μ_s'=1.99	μ_a= 0.05 μ_s'=1.29	μ_a= 0.0421 μ_s'=1.29
710nm	μ_a= 0.017 μ_s'=1.86	μ_a= 0.03 μ_s'=1.2	μ_a= 0.029 μ_s'=1.2

reconstruction with a Matrix-Free method. The reconstruction took 54sec. on a 2GHz Pentium with 2GB memory machine. The resulting reconstruction can be seen in the figure 6 on the right side, while slices along the center of the target and the reconstructed target can be found in figure 7.

5. Conclusions

In [11] we introduced a matrix-free method for multiwavelength reconstructions for Fluorescence Tomography that is able to deal with the large amount of data that are created by the non-

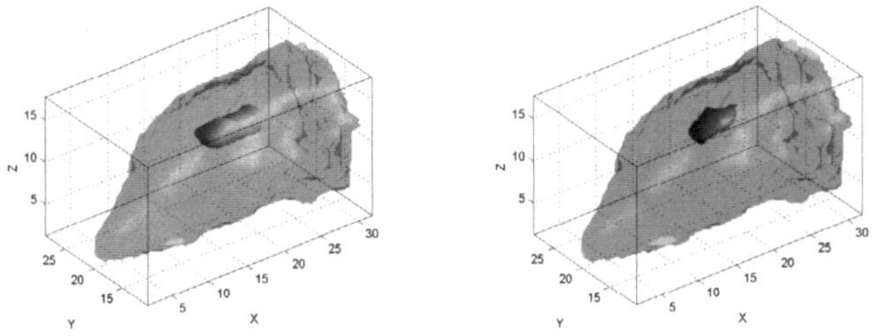

Figure 6. On the left the target rod used for the creation of the simulated data. On the right, the reconstructed rod from the simulated measurements with 5% gaussian noise. The reconstruction lasted 54sec.

Figure 7. Slices of the fluorescence concentration along the y (lower) and the z (top) axis along the slices that go through the center of the rod. On the left the target rod used for the creation of the simulated data.On the right the reconstructed rod from simulated data with added 5% noise. The reconstruction lasted 54sec.

contact acquisition and produce fast and good resolution reconstructions. Following that work, and having in-vivo applications in mind we tested the Matrix-Free method with in-vivo data from a mouse where fluorescence targets of different concentrations were inserted. Using the Matrix-free method we managed to reconstruct for the positions of the fluorescent tubes and a relative quantification accuracy and sensitivity. More specifically in figure 4, the algorithm demonstrated an almost linear response between the controlled concentrations and the reconstructed ones.

Apart from the slab-like reconstruction used for the in-vivo experiment and having in mind imaging applications in the brain, we also tested the algorithm with the use of realistic geometries, created from a mouse atlas. Anatomical and prior known optical properties were implemented in the forward model and demonstrated a reconstruction, figure 6 and figure 7, from simulated data which took 54sec to converge.

As future work we propose the use of a second modality like XCT of MRI to acquire a case specific mesh, and in the further extent the use of generic atlases to adopt a mesh to the specimens under investigation.

Acknowledgments

This work was funded by the EC Seventh Framework Grant, FMT-XCT, grant 201792 and EPSRC grant EP/E034950/1 .

References

[1] Ntziachristos V, Ripoll J, Wang V H L and Weissleder R (2005) Looking and listening to light: the evolution of whole body photonic imaging, *Nat Biotechnol.* **23**, 313–320 .

[2] Schulz R, Ripoll J, and Ntziachristos V (2003) Noncontact optical tomography of turbid media, *Opt. Lett.* **28**, 1701–1703.

[3] Ripoll J, Schultz R and Ntziachristos V, (2003) Free space propagation of diffuse light: theory and experiments, *Phys. Rev. Lett.* **91**, 103901.

[4] Meyer H, Garofalakis A, Zacharakis G, Psycharakis S, Mamalaki C, Kioussis D, Economou E N, Ntziachristos V and Ripoll J (2007) Noncontact optical imaging in mice with full angular coverage and automatic surface extraction, *Appl. Opt.* **46**, 3617–3627

[5] Culver J, Ntziachristos V, Holboke J M, and Yodh G A (2001) Optimization of optode arrangements for diffuse optical tomography: A singular-value analysis, *Opt. Lett.* **26**, 701–703

[6] Graves E, Ripoll J, Weissleder R and Ntziachristos V (2003) A submillimeter resolution fluorescence molecular imaging system for small animal imaging, *Med. Phys.* **30**, 901–911.

[7] Panasyuk G Y, Wang Z M, Schotland J C and Markel V A (2008) Fluorescent optical tomography with large data sets, *Opt. Lett.* **33**, 1744–1746 .

[8] Wang Z M, Panasyuk G Y, Markel V A and Schotland J C (2005) Experimental demonstration of an analytic method for image reconstruction in optical diffusion tomography with large data sets, *Opt. Lett.* **30**, 3338–3340 .

[9] Ripoll J (2010) Hybrid Fourier-real space method for diffuse optical tomography, *Opt. Lett.* **35**, 688–690

[10] Rudge J T, Soloviev Y V and Arridge R S (2010) Fast image reconstruction in fluoresence optical tomography using data compression, *Opt. Lett.* **35**, 763–765

[11] Zacharopoulos A, Svenmarker P, Axelsson J, Schweiger M, Arridge S and Andersson-Engels S (2009) A matrix-free algorithm for multiple wavelength fluorescence tomography, *Opt. Express* **17**, 3042–3051.

[12] Arridge R S (1999) Optical tomography in medical imaging, *Inverse Problems* **15**, 41–93 .

[13] Bourayou R, Boeth H, Benav H, Betz T, Lindauer U, Nierhaus T, Klohs J, Wunder A, Dirnagl U and Steinbrink J (2008) Fluorescence tomography technique optimized for noninvasive imaging of the mouse brain. *J Biomed Opt* **13** (4) 041311

[14] McCann C M, Waterman P, Figueiredo J-L, Aikawa E, Weissleder R and Chen J W (2009) Combined magnetic resonance and fluorescence imaging of the living mouse brain reveals glioma response to chemotherapy. *Neuroimage*: **45**(2) 360-369.

[15] Hyde D, Schulz R, Brooks D, Miller E and Ntziachristos V (2009) Performance dependence of hybrid x-ray computed tomography/fluorescence molecular tomography on the optical forward problem. *J Opt Soc Am A Opt Image Sci Vis* **26**(4) 919–23.

[16] Hyde D, Kleine R de, MacLaurin S A, Miller E, Brooks D H, Krucker T and Ntziachristos V (2009) Hybrid FMT-CT imaging of amyloid-[beta] plaques in a murine Alzheimer's disease model. *Neuroimage* **44**(4): 1304–1311.

[17] Dogdas B, Stout D, Chatziioannou A, and Leahy R M (2007) Digimouse: A 3D Whole Body Mouse Atlas from CT and Cryosection Data. *Phys Med Biol.* **52** (3): 577–87

[18] Stout D, Chow P, Silverman R, Leahy R M, Lewis X, Gambhir S and Chatziioannou A (2002) Creating a whole body digital mouse atlas with PET, CT and cryosection images. *Mol. Imag. and Biol.* **4** (4): S27.

[19] Fang Q, Carp S A, Selb J, Moore R, Kopans D B, Miller E L, Brooks D H, and Boas D A (2008) A Multi-Modality Image Reconstruction Platform for Diffuse Optical Tomography, *Biomedical Optics*, BIOMED Poster Session II (BMD)

[20] Ntziachristos V and Weissleder R (2001) Experimental three-dimensional fluorescence reconstruction of diffuse media by use of a normalized Born approximation, *Opt. Lett.* **26**, 893-895

[21] Garofalakis A, Zacharakis G, Meyer H, Economou E N, Mamalaki C, Papamatheakis J, Kioussis D, Ntziachristos V and Ripoll J (2007) 3D in-vivo imaging of GFP-expressing T-cells in Mice with non-contact Fluorescence Molecular Tomography, *Mol Imaging.* **6**, 96–107

[22] Martin A, Aguirre J, Sarasa A, Garofalakis A, Meyer H, Mamalaki C, Ripoll J and Planas A M (2008) Non-invasive in vivo assessment of stroke-induced immune depression by three-dimensional fluorescence molecular tomography in transgenic mice expressing GFP in T-lymphocytes, *Mol. Imag.*, **7**, 157–67

[23] Lin Y, Gao H, Nalcioglu O, and Gulsen G (2007) Fluorescence diffuse optical tomography with functional and anatomical a priori information: feasibility study, *Phys. Med. Biol.*, **52**, 5569–5585

AUTHOR INDEX

Institute of Physics
Dirac House, Temple Back
Bristol BS1 6BE UK

CURRAN ASSOCIATES INC
proceedings.com